Engineering and Technology

Today engineers have available personal microcomputers to assist them in analysis and design. This is a great advantage over the capability that existed previously.

Careers
in Engineering and Technology

THIRD EDITION

George C. Beakley, P.E.

Professor of Engineering, Arizona State University

Macmillan Publishing Company

NEW YORK

Collier Macmillan Publishers

LONDON

Copyright © 1984, *Macmillan Publishing Company,*
a division of Macmillan, Inc.

Printed in the United States of America

Earlier editions entitled *Elementary Problems in Engineering*
copyright 1949 and 1951 by H. W. Leach and George C.
Beakley; *Engineering: The Profession and Elementary
Problem Analysis,* Chapters 10 and 11, Preliminary Edition,
© 1959 by H. W. Leach and George C. Beakley;
*Engineering: The Profession and Elementary Problem
Analysis* © 1960 by H. W. Leach and George C. Beakley.
Material reprinted from *Engineering: An Introduction to a
Creative Profession* © copyright 1967 by George C.
Beakley and H. W. Leach. Copyright © 1969 by Macmillan
Publishing Co., Inc.

Macmillan Publishing Company
866 Third Avenue, New York, New York 10022

Collier Macmillan Canada, Inc.

Library of Congress Cataloging in Publication Data

Beakley, George C.
   Careers in engineering and technology.

   Includes bibliographical references and index.
   1. Engineering—Vocational guidance.   2. Technology—Vocational guidance.   I. Title.
TA157.B39 1984   620′.0023   84-868
ISBN 0-02-307110-9

Printing: 12345678   Year: 456789012

ISBN   0-02-307110-9

Preface

Technology is changing at an ever-increasing pace. Never before have so many compelling technological problems occupied positions of prominence in man's system of values. Today the entering college student is offered the opportunity to prepare for entrance into a variety of professional programs. Among the more important of these is engineering and technology, because those who make this choice will help to shape the destiny of civilizations yet unborn. Challenge and opportunity? Yes . . . but not without a consciousness and a realization of the moral and ethical responsibilities that should accompany the emergence of new processes, designs, products, and systems. Recognizing the importance of an early commitment toward these ends, this work has been prepared as an informational and motivational instrument for the use of those who are interested in preparing themselves today for the solution of tomorrow's problems.

This third edition continues the multipurpose format of the other editions. It is designed for use in (1) informational courses that introduce the student to careers within the broad engineering spectrum, (2) courses with an analysis or systems emphasis, and (3) courses that introduce the student to both the excitement and frustration of engineering design. A number of teachers have preferred to organize introductory courses that draw material from each of these areas. In preparing this edition the author has also incorporated a number of the features of his more complete work, *Engineering: An Introduction to a Creative Profession,* Macmillan (4th edition).

This edition has increased the options available to the teacher and student, and has placed increased emphasis on the expanding role of the technician and technologist in modern industry. Several new chapters have been added, including a comprehensive chapter on the modeling of engineering systems—mechanical, electrical, fluid, and thermal. This elementary multidisciplinary treatment gives the student an insight into the analysis of engineering systems. It emphasizes the commonality that exists among the engineering disciplines. However, in mathematical rigor this chapter requires only the use of algebra. Many engineering and technology schools will require study of this same material at the junior level, but using calculus as a mathematical base. Another new feature is the addition of Chapter 10, "Designs in Nature." Here the author develops student interest in engineering design by describing a number of the ingenious designs of nature and by showing how these have been adapted to our use. Another new and important feature is Chapter 8, "The Use of Computers in Engineering and Technology." This chapter is informational in style—rather than computational. It summarizes the opportunities in analysis and design made available to the student by the advent of computer systems. Particular consideration is given to enhanced graphics capability and the use of the microcomputer for elementary design work. Appendix I contains a comprehensive review of the fundamentals of technical drawing, which is especially useful in conjunction with teaching the principles of engineering design. It can also serve as an instructional basis for an abbreviated study of the principles of technical drawing.

In preparing this book the author has been aided not only by reviews and criticisms of the previous edition, but also by comments and suggestions from numerous professors and practicing engineers and technologists in industry throughout the United States. Special appreciation is expressed to John Michael Sims, who made several significant contributions to Chapter 8, particularly with regard to information pertaining to the use of microcomputers. A number of colleagues of the author have also been most influential in giving this book a unique blend of the academic and industrial viewpoints.

The author wishes to thank those who have offered advice and encouragement and those who have made recommendations as to format and content for this new third edition. He also gratefully acknowledges the assistance of the many industries who have supplied illustrative materials for inclusion in the text.

George C. Beakley

Contents

Acknowledgments

Frontispiece	Radio Shack, a Division of the Tandy Corporation	Illustration 2-7	Jervis B. Webb Company
Illustration 1-2	Maddox and Hopkins	Illustration 3-1	North American Rockwell
Illustration 1-3	W. S. Dickey Clay Manufacturing Company	Illustration 3-2	International Harvester Company
Illustration 1-5	Maddox and Hopkins	Illustration 3-3	Aluminum Company of America
Illustration 1-6	Ewing Galloway	Illustration 3-4	Aerojet-General Corporation
Illustration 1-7	Maddox and Hopkins		
Illustration 1-8	Maddox and Hopkins	Illustration 3-5	Glass Container Manufacturers Institute
Illustration 1-12	Fisher Scientific Company		
Illustration 1-14	Texas Instruments Incorporated	Illustration 3-6	General Electric Research and Development Center
Illustration 1-15	General Electric FORUM	Illustration 3-7	Bethlehem Steel Corporation
Illustration 1-16	National Air and Space Museum, Smithsonian Institute	Illustration 3-8	Intel Corporation
		Illustration 3-9	RCA Electronic Components
Illustration 1-17	Bell Laboratories		
Illustration 1-18	Matheson Gas Products	Illustration 3-10	General Electric Research and Development Center
Illustration 1-19	Battelle Development Corporation		
		Illustration 3-11	Union Carbide Corporation
Illustration 1-20	National Aeronautics and Space Administration	Illustration 3-12	Cities Service Company
		Illustration 3-13	General Dynamics Electric Boat Division
Illustration 1-21	Rockwell International Corporation/Space Division		
		Illustration 3-14	Los Alamos Scientific Laboratory
Illustration 2-1	Shell Oil Company		
Illustration 2-2	Hughes Aircraft Company	Illustration 3-15	Humble Oil and Refining Company
Illustration 2-3	Ford Motor Company		
Illustration 2-4	The Boeing Corporation	Illustration 4-1	Dr. J. E. Cermack, Director, Fluid Dynamics and Diffusion Laboratory, Colorado State University
Illustration 2-5	Westinghouse Electric Corporation		
Illustration 2-6	Union Carbide Corporation		

Illustration 4-2   Monsanto Chemical Company

Illustration 4-3   General Motors Corporation

Illustration 4-5   *Industrial Engineering*

Illustration 4-6   Beech Aircraft Corporation

Illustration 4-7   Bethlehem Steel Corporation

Illustration 4-8   The Upjohn Company

Illustration 4-9   Jervis B. Webb Company

Illustration 4-10   Fisher Scientific Company

Illustration 5-1   Ernst & Ernst

Illustration 5-3   General Dynamics

Illustration 5-4   Floyd Clark, California Institute of Technology

Illustration 5-5   *Design News,* A Cahners Publication

Illustration 6-1   Information Handling Services

Illustration 6-2   Chicago Bridge & Iron Company

Illustration 6-3   Rockwell International

Illustration 7-1   Acme-Cleveland Corporation

Illustration 7-2   Radio Shack, a Division of the Tandy Corporation

Illustration 7-3   Acme-Cleveland Corporation

Illustration 7-5   Walker/Parkersburg

Illustration 7-6   Walker/Parkersburg

Illustration 7-7   National Aeronautics and Space Administration

Illustration 7-8   Somerset Importers, Ltd.

Figure 8-1   Honeywell Information Systems, Inc.

Figure 8-2   Intergraph Corporation

Figure 8-3   Bausch & Lomb

Figure 8-4   Computervision Corporation

Figure 8-5   Bausch & Lomb

Figure 8-6   Computervision Corporation

Figure 8-7   Bausch & Lomb

Figure 8-8   Evans & Southerland Computer Corporation

Figure 8-9   IBM Corporation

Figure 8-10   Apple Computer, Inc.

Figure 8-11   IBM Corporation

Figure 8-12   Honeywell Information Systems, Inc.

Figure 8-13   Aristographics Corporation

Figure 8-14   Hewlett-Packard Company

Figure 8-15   Computervision Corporation

Figure 8-16   Computervision Corporation

Figure 8-17   Bausch & Lomb

Figure 8-18   Computervision Corporation

Figure 8-19   Computervision Corporation

Figure 8-20   Computervision Corporation

Figure 10-1   Stennett Heaton Photo, Courtesy Neil A. Maclean Co., Inc.

Figure 10-2   Moody Institute of Science

Illustration 10-3   Photo by Gordon Smith from National Audubon Society

Illustration 10-4   RCA Corporation

Illustration 10-5   Photo by John H. Gerard from National Audubon Society

Illustration 10-6   General Dynamics, Convair Division

Illustration 10-7   Carolina Biological Supply Company

Illustration 10-9   U.S. Department of Agriculture

Illustration 10-10   Photo by Treat Davidson from National Audubon Society

Illustration 10-11   Photo by Robert C. Hermes from National Audubon Society

Illustration 10-12   Moody Institute of Science

Illustration 10-13   Moody Institute of Science

Illustration 10-14   Photo by G. E. Kirkpatrick from National Audubon Society

Illustration 10-15   Carl Zeiss Inc., New York

Illustration 10-16   American Iron and Steel Institute

Figure 11-1   Robert Shaw Controls Company

Illustration 11-2   First National Bank of Arizona

Illustration 11-3   United Aircraft

Illustration 11-4   General Motors Corporation and American Society of Civil Engineers

Illustration 11-5   Allis-Chalmers

Illustration 11-6   *Kaiser News*

Illustration 11-11   United States Navy

Illustration 11-12   Can-Tex Industries

Illustration 11-13   Exxon Chemical Company, USA

Illustration 11-14   Encyclopaedia Britannica

Illustration 11-15   The Falk Corporation

Illustration 11-16   Al Capp

Illustration 11-17   Chesebrough-Pond's Inc.

Illustration 11-18   New York Life Insurance Co.

Illustration 11-19   Frank Roberge and *Arizona Republic*

Illustration 11-21   Arizona State University

Illustration 11-22   National Aeronautics and Space Administration

Figure 11-10   General Dynamics, Electric Boat Division

Illustration 11-25   Texas Instruments Incorporated

Illustration 11-27   Fisher Scientific Company

Illustration 11-28   Reynolds Aluminum, Reynolds Metals Company

Illustration 11-29   Ford Motor Company

Illustration 11-30	Rocketdyne, North American Rockwell	A3	Keuffel & Esser Co.
Illustration 11-31	RCA Corporation	A4a	Gramercy (Compass and Accessories)
Illustration 12-1	Arizona Republic	A4b	Keuffel & Esser Co. (Circle Template, Irregular Curve)
Illustration 12-2	Ford Motor Company Design Center	A5a	J. S. Staedtler, Inc. (Divider)
Illustration 12-3	Midas, Inc.	A5b	Keuffel & Esser Co. (Scale)
Illustration 12-4	Texas Instruments Incorporated	A6	J. S. Staedtler, Inc.
		A7	Berol USA
Illustration 12-5	Enjay Chemical	A9a	J. S. Staedtler, Inc. (Rapidograph Pens)
Illustration 12-6	Phoenix Newspapers, Inc.		
Appendixes		A9b	Koh-I-Noor Rapidograph, Inc. (Compass)
Part Opening	Sandia Laboratories		
Ala	J. S. Staedtler, Inc. (T-square)	A10	J. S. Staedtler, Inc.
		A35	Koh-I-Noor Rapidograph, Inc.
Alb	Martin Instrument Co. (Parallel Rule)	A36	Keuffel & Esser Co.
A2	Martin Instrument Co.		

Careers in Engineering and Technology

1

Engineering in history

When did engineering begin? Who were the first engineers? What were the objectives of work by the early engineers? Answers to these questions and others concerning the beginning of engineering appear in the fragments of historical information available to us. In fact, the beginnings of civilization and the beginnings of engineering are coincident. As early man emerged from caves to make homes in communities, he adapted rocks and sticks as tools to aid him. Simple as these items may seem to us today, their useful employment suggests that the creative ideas which emerged in the minds of early man were developed into useful products to serve the recognized needs of the day. Some served as tools in the struggle for existence of an individual or group, and others were used for protection against wild animals or warlike neighbors. (Illustration 1-1). Early engineering was therefore principally either civil or military.

Down through the ages, the engineer has been in the forefront as a maker of history. Material accomplishments have had as much impact on world history as any political, economic, or social development. Sometimes these accomplishments have stemmed from the pressures of need from evolving civilizations. At other times abilities to produce and meet needs have led the way for civilizations to advance. In general, engineers do the things required to serve the needs of the people and their culture.

Basically, the role of the engineer has not changed through the centuries. The primary task has always been to take knowledge and make practical use of it—converting scientific theory into useful application, and in so doing, providing for mankind's material needs and well-being. From era to era, only the objectives pursued, the techniques of solution used, and the available tools of analysis have changed.

It is helpful to review the past to gain insight to the driving forces of science and to learn of the individuals who developed and applied these principles. A review also

Illustration 1-1
Primitive man fashioned tools to sustain his existence.

will reveal certain facts concerning the discovery and use of fundamental scientific principles. Primarily, science builds its store of knowledge on facts which, once determined, are available from then on for further discovery. This principle is in contrast to the arts, since, for example, the ability of one person to produce a beautiful painting does not make available to others his skills in producing paintings.

Outstanding characteristics of engineers through the centuries have been a willingness to work and an intellectual curiosity about the behavior of things. Their queries about "Why?," "How?," "With what?," and "At what cost?" have all served to stimulate an effort to find desirable answers to many types of technological problems.

Another characteristic associated with engineers is the ability to "see ahead." The engineer must have a fertile imagination, must be creative, and must be ready to accept new ideas. Whether an engineer lived at the time of construction of the pyramids or has only recently graduated in nuclear engineering, these characteristics have been an important part of his intellectual makeup.

The following sections present a brief picture of the development of engineering since the dawn of history and outline the place that the engineer has held in various civilizations.

The beginnings of engineering: 6000 B.C.–3000 B.C.

The beginning of engineering probably occurred in Asia Minor or Africa some 8000 years ago. About this time, man began to cultivate plants, domesticate animals, and build permanent houses in community groups. With the change from a nomadic life came requirements for increased food production. Among the first major engineering projects were irrigation systems to promote crop growing. Increased food production permitted time for men to engage in other activities. Some became rulers, some priests, and many became artisans, whom we may call the first engineers.

Early achievements in this era included methods of producing fire at will, melting certain rocklike materials to produce copper and bronze tools, invention of the wheel and axle, development of a system of symbols for written communication, origination of a system of mathematics, and construction of irrigation works.

Early records are so fragmentary that only approximate dates can be given for any specific discovery, but evidence of the impact of early engineering achievements is readily discernible. For example, in setting up stable community life in which land was owned, there had to be provision both for irrigation and for accurate location and maintenance of boundaries. This necessity stimulated the development of surveying and of mathematics. The moving of earth to make canals and dams required computations, and to complete the work the efforts of many men had to be organized and directed. As a result, a system of supervisors, foremen, and workers was established that formed the beginnings of a class society.

In this society, craftsmen became a distinct group producing useful items such as pottery, tools, and ornaments that were desired by others. As a result, trade and commerce were stimulated and roads were improved. Some 5000 years ago man first used the wheel and axle to make two-wheeled carts drawn by animals.

In order to record the growing accumulation of knowledge about mathematics and engineering, the early engineer needed a system of writing and some type of writing material. In the Mesopotamian region, soft clay was used on which cuneiform characters were incised. When baked, the clay tile material was used for permanent documents, some of which are legible even today (Illustration 1-2). In the Nile Valley, a paperlike material called *papyrus* was made from the inner fibers of a reed. In other parts of Asia Minor, treated skins of animals were used to form parchment. Occasionally, slabs of stone or wood were used as writing materials. The type of writing that developed was strongly influenced by the writing material available. For example, the incised characters in soft clay differed significantly from the brush strokes used in writing on papyrus (Illustration 1-4).

In engineering work, a source of energy is necessary. This requirement led to the enslavement and use of numbers of human beings as primary sources of energy. The construction of all early engineering works, whether Oriental, Mediterranean, or American Indian, was accomplished principally by human labor. It was not until near the end of the period of history known as the Middle Ages that mechanical sources of power were developed.

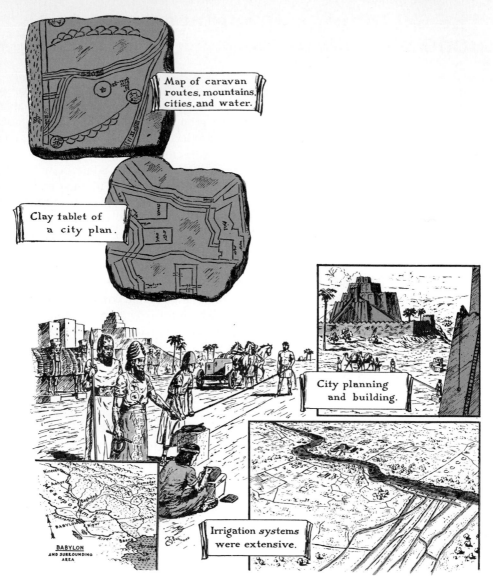

Mesopotamia, often called the "Cradle of civilization," could also be said to have nurtured engineering in its infancy. Clay tablets, such as the ones shown on this page, have been unearthed which show city plans, irrigation, water supply systems, and what appear to be road maps. Although no engineering tools have been discovered among the remains of ancient Mesopotamia, the evidence unearthed of their remarkable architectural construction indicates that they used measuring tools, which, even though primitive, aided in producing engineering of a high degree for this period. Their cities, with their water supply, irrigation systems, and road networks, were among the wonders of the ancient world.

Many outstanding contributions of mathematics were made by the Mesopotamians. It has been proven that they had knowledge of the sexagesimal system, in which they divided the circle into 360 degrees, the hour into 60 minutes and the minute into 60 seconds.

Illustration 1-2

Mesopotamia, often called the cradle of civilization, also may be said to have begun engineering. Excavations have revealed extensive architecture, irrigation systems, roads, and land planning. In this picture is shown a party of surveyors using tools for measurement which, for the period, were remarkably accurate.

Illustration 1-3
Ancient builders employed engineering principles in the construction of their structures. Clay plumb bobs, such as the one pictured here being used by Babylonian builders, have been unearthed recently by archaeologists.

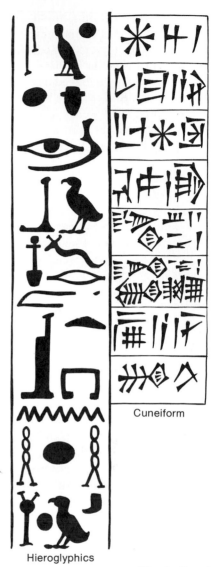

Cuneiform

Hieroglyphics

Illustration 1-4

Engineering in early civilizations: 3000 B.C.–600 B.C.

After about 3000 B.C., enough records were made on clay tablets, on papyrus and parchment, on pottery, and as inscriptions on monuments and temples to provide us with information about ancient civilization. These records show that urban civilizations existed in Egypt, Mesopotamia, and the Indus Valley, and that a class society of craftsmen, merchants, soldiers, and government officials was a definite part of that civilization.

In Mesopotamia, clay tablets have been uncovered which show that Babylonian engineers were familiar with basic arithmetic and algebra. From these writings we know that they routinely computed areas and volumes of land excavations. Their number system, based on 60 instead of 10, has been handed down through the centuries to us in our measures of time and angle. Their buildings were constructed principally of baked brick. Primitive arches were used in some of their early hydraulic works. Bridges were built with stone piers carrying wooden stringers for the roadway. Some roads were surfaced with a naturally occurring asphalt, a construction method that was not used again until the nineteenth century.

It was in Egypt that some of the world's most remarkable engineering was performed (Illustration 1-5). Beginning about 3000 B.C. and lasting for about 1000 years, the Pyramid Age flourished in Egypt. The first pyramids were mounds covered with stone, but the techniques progressed rapidly until the Great Pyramid was begun about 2900 B.C. Stones for the structures were cut by workmen laboriously chipping channels in the native rock, using a ball made of a harder rock as a tool. By this method, blocks weighing 15 tons or more were cut for use in building. Over 2,300,000 building stones weighing about 5000 pounds each were used to construct the Great Pyramid of Cheops. It stands 481 feet high. The Egyptian engineers apparently used only the lever, the inclined plane, the wedge, and the wheel in their construction efforts (Illustration 1-6).

Although early construction tools were primitive, the actual structures, even by today's standards, are outstanding examples of engineering skill in measurement and layout. For example, the base of the Great Pyramid is square within about 1 inch in a distance of 756 feet, and its angles are in error by only a few minutes despite the fact that the structure was built on a sloping rocky ledge.

The Egyptian engineers and architects held a high place in the Pharaoh's court. Imhotep, a designer of one of the large pyramids, was so revered for his wisdom and ability that he was included as one of the Egyptian gods after his death. Not only were the Egyptian engineers skilled builders, they were also skilled in land measurement. Annual overflows of the Nile River obliterated many property lines and a resurvey of the valley was frequently necessary. Using geometry and primitive measuring equipment, they restored markers for land boundaries after the floods receded.

The Egyptians also were skilled in irrigation work. Using a system of dikes and canals, they reclaimed a considerable area of desert. An ancient engineering contract to build a system of dikes about 50 miles long has recently been discovered.

Although the skill and ingenuity of the Egyptian engineers were outstanding, the culture lasted only a relatively short time. Reasons which may account for the failure to maintain leadership are many, but most important was the lack of pressure to continue development. Once the engineers formed the ruling class, little influence could be brought to bear to cause them to continue their creative efforts. Since living conditions were favorable after an agricultural system was established, little additional engineering was required. The lack of urgency to do better finally stifled most of the creativity of the engineers and the civilization fell into decay.

In ancient Egypt warfare and strife delayed the development of engineering; however, with the unification of Upper and Lower Egypt, the science of measurement and construction made rapid progress. Buildings, city planning, and irrigation systems show evidence of this development. Good judgment and reasonable engineering design resulted in sound and durable structures. The Pyramids are engineering marvels both in design and construction.

That the Egyptians advanced mathematics is attested to by papyrus scrolls, dating back to 1500 B.C., which show that the Egyptians had knowledge of the triangle and were able to compute areas and volumes. They also had a device to obtain the azimuth from the stars.

The annual floods of the Nile afforded ample practice in measurement surveying. This may well have been the first example of the importance of resurveys. The rope used as a measure was first soaked in water, dried, and then coated heavily with wax to ensure constant length. Probably some crude surveying instruments were devised, but none have been found.

Illustration 1-5

In Egypt the science of measurement and construction developed rapidly. The pyramids are engineering marvels both in design and construction. Papyrus scrolls show that the Egyptians had knowledge of the triangle and were able to compute areas and volumes.

Illustration 1-6
The pyramids of Egypt exemplify man's desire to create and build enduring monuments.

Science of the Greeks and Romans: 600 B.C.—A.D. 400

The history of engineering in Greece had its origins in Egypt and the East. With the decline of the Egyptian civilization, the center of learning shifted to the island of Crete and then about 1400 B.C. to the ancient city of Mycenae in Greece.

To the engineers of Mycenae were passed not only the scientific discoveries of the Egyptians but also a knowledge of structural building materials and a language that formed the basis of the early Greek language. These engineers subsequently developed the corbeled arch and made wide use of irrigation systems.

The Greeks of Athens and Sparta borrowed many of their developments from the Mycenaean engineers. In fact, the engineers of this period were better known for the intensive development of borrowed ideas than for creativity and invention. Their water system, for example, modeled after Egyptian irrigation systems, showed outstanding skill in the use of labor and materials, and these Greeks established technical procedures that have endured for centuries (Illustration 1-7).

Greece was famous for its outstanding philosophers. Significant contributions

Illustration 1-7
The Greeks constructed many buildings of unusual beauty which show a high degree of engineering skill and architectural design. Their cities had municipal water supplies that required dams and aqueducts to bring water from the mountains. This picture shows a builder laying out a building foundation, using a divided circle, a plumb bob, and a knotted rope.

Hydraulics provide public water

Aqueducts, tunnels and highways

The outstanding progress made by the Ancient Grecians in architecture and mathematics and their contribution to the advancement of engineering demand our admiration.

Aristotle contended that the world was a spheroid. He stated that observations of the various stars showed the circumference of the earth to be about 400,000 *stadia* (4600 miles).

Erathosthenes of Cyrene observed that the sun's rays, when perpendicular to a well at Alexandria, cast a shadow equal to one fiftieth of a circle at Syene (Aswan) 500 miles away. Thus he established that the circumference of the earth was 50 times 500 miles or 25,000 miles.

The Greeks constructed many buildings and structures of large size, which show engineering skill and excellent architectural design. One tunnel, which was built to bring water to Athens, measured 8 feet by 8 feet and was 4200 feet in length. The construction of such a tunnel necessitated extremely accurate alignment both on the surface and underground.

> When looms weave by themselves man's slavery will end.
> —Aristotle

were made by men such as Plato, Aristotle, and Archimedes. In the realm of abstract thought, they perhaps have never been equaled, but at that time extensive use of their ideas was retarded because of the belief that verification and experimentation, which required manual labor, were fit only for slaves. Of all the contributions of the Greeks to the realm of science, perhaps the greatest was the discovery that nature has general laws of behavior which can be described with words.

The Great Wall of China, over 1500 miles long, was completed in the third century B.C. It averaged 25 feet in height, 15 feet in width at the top, and 25 feet in width at the base.

The best engineers of antiquity were the Romans. Within a century after the death of Alexander, Rome had conquered many of the eastern Mediterranean countries, including Greece. Within two more centuries Rome had dominion over most of the known civilized areas of Europe, Africa, and the Middle East. Roman engineers liberally borrowed scientific and engineering knowledge from the conquered countries for use in warfare and in their public works. Although in many instances they lacked originality of thought, Roman engineers were superior in the application of techniques (Illustration 1-8).

From experience Rome had learned the necessity for establishing and maintaining a system of communications to hold together the great empire (Illustration 1-9). Thus Roman roads became models of engineering skills. By first preparing a deep subbase and then a compact base, the Romans advanced the technique of road construction so far that some Roman roads are still in use today. At the peak of Roman sovereignty, the network of roads comprised over 180,000 miles stretching from the Euphrates Valley to Great Britain.

In addition, Roman engineers were famous for the construction of aqueducts and bridges. Using stone blocks in the constructing of arches, they exhibited unusual skill. An outstanding example of this construction is the famous Pont du Gard near Nîmes, France, which is 150 feet high and over 900 feet long. It carries both an aqueduct and a roadway.

By the time of the Christian era, iron refining had developed to the extent that iron was being used for small tools and weapons. However, the smelting process was so inefficient that over half of the metallic iron was lost in the slag. Except in the realm of medicine, no interest was being shown in any phase of chemistry.

Despite their outstanding employment of construction and management techniques, the Roman engineers seemed to lack the creative spark and imagination necessary to provide the improved scientific processes required to keep pace with the expanding demands of a far-flung empire. The Romans excelled in law and civil administration but were never able to bring distant colonies fully into the empire. Finally, discontent and disorganization within the empire led to the fall of Rome to a far less cultured invader.

Illustration 1-8

The rise of the Roman Empire was attributed to the application of engineering principles to military tactics. This picture shows a construction party as they build a section of the famous Roman highways. Notice the heavy foundations which exist to this day.

Scientific approach to navigational problems.

Piers and arches, a product of geometry

ROME
AND PART OF THE
ROMAN EMPIRE
AT ITS HEIGHT

The Romans excelled in the building of aqueducts. Many of these carried water for great distances with perfect grade and alignment. The key design in this type of construction was the arch, which was also used in bridges, tunnels, buildings, and other construction.

Evidence of the Romans' knowledge and understanding of basic geometric principles is further shown by their river and harbor construction and the scientific approach to navigational problems.

Sanitary systems, paved roads, magnificent public buildings, water supply systems, and other public works still in evidence today stand as monuments to the Roman development of engineering as a key to the raising of the standard of living.

The rise of the Roman Empire was attributable to the application of engineering principles applied to military tactics. The invincibility of the Roman legions was the result not only of the valor of the fighting men but also, and perhaps more strongly, to the genius of the Roman military engineers.

Illustration 1-9

". . . The day's march is complete, centurion . . ." The Roman army used a gear driven odometer on the axle of the unit commander's chariot to measure distance traveled. For each measured distance a round stone was discharged into a cup on the tail of the chariot.

Engineering in the middle ages: fifth to sixteenth centuries

After the fall of Rome, scientific knowledge was dispersed among small groups, principally under the control of religious orders. In the East, an awakening of technology began among the Arabs but little organized effort was made to carry out any scientific work. Rather, it was a period in which isolated individuals made new discoveries or rediscovered earlier known scientific facts.

It was during this time that the name *engineer* first was used. Historical writings of about A.D. 200 tell of an *ingenium,* an invention, which was a sort of battering ram used in attacks on walled defenses. Some thousand years later, we find that an *ingeniator* was the man who operated such a device of war—the beginning of our modern title, *engineer.*

Several technical advances were made late in this period. One important discovery involved the use of charcoal and a suitable air blast for the efficient smelting of iron. Another advance was made when the Arabs began to trade with China and a process of making paper was secured from the Chinese. Within a few years the Arabs had established a paper mill and were making paper in large quantities. With the advent of paper, communication of ideas began to be reestablished. Also in Arabia, the sciences of chemistry and optics began to develop. Sugar refining, soap making, and perfume distilling became a part of the culture. The development of a method of making gun powder, probably first learned from China about the fourteenth century, also had rapid and far-reaching results (Illustration 1-10).

Illustration 1-10
*The hand cannon
ended the superiority of
armor.*

Illustration 1-11
Leonardo da Vinci . . . artist, architect, engineer . . . a creative genius for all ages and all seasons.

After centuries of inaction, the exploration of faraway places began again, aided greatly by the development of a better compass. With the discovery of other cultures and the uniting of ideas, there gradually emerged a reawakening of scientific thought.

With the growth of Christianity, an aversion arose to the widespread use of slaves as primary sources of power. This led to the development of waterwheels and windmills and to a wider use of animals, particularly horses, as power sources.

About 1454, Gutenberg, using movable type, produced the first books printed on paper. This meant that the knowledge of the ages, which previously had been recorded laboriously by hand, now could be disseminated widely and in great quantities. Knowledge, which formerly was available only to a few, now was spread to scholars everywhere. Thus the invention of paper and the development of printing served as fitting climaxes to the Middle Ages.

Seldom has the world been blessed with a genius such as that of Leonardo da Vinci (1452–1519). Although still acclaimed today as one of the greatest of all artists, his efforts as an engineer, inventor, and architect are even more impressive. Long after his death his designs of a steam engine, machine guns, a camera, conical shells, a submarine, and a helicopter have been proven to be workable (Illustration 1-11).

Galileo (1564–1642) was also a man of great versatility. He was an excellent writer, artist, and musician, and he is also considered one of the foremost scientists of that period. One of his greatest contributions was his formulation of what he considered to be the scientific method of gaining knowledge.

The revival of science: seventeenth and eighteenth centuries

Following the invention of printing, the self-centered medieval world changed rapidly. At first, the efforts to present discoveries of Nature's laws met with opposition and in some cases even hostility. Slowly, however, freedom of thought was permitted and a new concept of *testing to evaluate a hypothesis* replaced the early method of establishing a principle solely by argument.

Four men in this period made discoveries and formulated laws which have proved to be of great value to engineering. They were Boyle, who formulated a law relating pressures and volumes of gases; Huygens, who investigated the effects of gravitational pull; Hooke, who experimented with the elastic properties of materials; and Newton, who is famous for his three laws of motion. All the early experimenters were hampered by a lack of a concise vocabulary to express their ideas. Because of this many of the principles were expressed in a maze of wordy statements.

During this period, significant advancements were made in communication and transportation. Canals and locks were built for inland water travel and docks and harbors were improved for ocean commerce. Advances in ship design and improved methods of navigation permitted a wide spreading of knowledge that formerly had been isolated in certain places.

The search for power sources to replace human labor continued. Water power and wind power were prime sources, but animals began to be used more and more. About this time, the first attempts to produce a steam engine were made by Papin and Newcomen. Although these early engines were very inefficient, they did mark the beginning of power from heat engines.

An important industry was made possible in this period by the development of spinning and weaving machinery by such men as Jurgen, Hargreaves, Crampton, and Arkwright. This period also marked a general awakening of science after the Dark Ages. Individual discoveries, although usually isolated, found their way into useful products within a short period of time because of the development of printing and the improvements in communication.

The basic discoveries in this era were made by men who were able to reject old, erroneous concepts and search for principles that were more nearly in accord with Nature's behavior. Engineers in any age must be equally discerning if their civilization is to advance.

Beginnings of modern science: nineteenth century

Early in the nineteenth century, two developments provided an impetus for further technological discoveries. The two developments were the introduction of a method, developed by Henry Cort, of refining iron and the invention of an efficient steam engine by James Watt (Illustration 1-12). These developments provided a source of iron for machinery and power plants to operate the machinery.

As transportation systems began to develop, both by water and by land, a network of railroads and highways was built to tie together the major cities in Europe and in the United States.

In this period, the awakening of science and engineering truly had begun. Now, although people were slow to accept new ideas, knowledge was not rejected as it had been in earlier centuries. Colleges began to teach more and more courses in science and engineering, and it was here that the fuse was lighted for an explosion of discoveries in the twentieth century.

One of the most important reasons for the significant development of technology in this period was the increasingly close cooperation between science and engineering. It began to become more and more evident that discoveries by research scientists could be used to develop new articles for commerce. Industry soon began to realize that money spent for research and development eventually returned many times its value.

Illustration 1-12
This engraving of James Watt, a Scotch engineer, first appeared in 1860. It portrays an experimental design which was to become the forerunner of the modern condensing steam engine. Watt had been given a Newcomen engine to repair, in 1764, and, noting its extreme wastefulness of fuel, set about the task of building a better machine.

Illustration 1-13
Colt's "Walker" pistol. The first military pistol made on an "assembly line" with interchangeable parts.

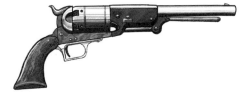

Twentieth-century technology

As the twentieth century came into being, a number of inventions emerged that were destined to have far-reaching effects on our civilization. The automobile began to be more widely used as better roads were made available. In the early part of the twentieth century Henry Ford began to build and sell automobiles, and he provided American industry with a new dimension in manufacturing. More than any other person he has been universally acclaimed as the individual responsible for bringing modern mass production into being. Today much of industry follows his industrial philosophy—to reduce the price of the product while improving its quality, to in-

A MANUFACTURING LANDMARK
FORD MECHANIZATION, 1915

Illustration 1-14
Henry Ford's first assembly line of 1915 is acclaimed today as a milestone toward modern manufacturing practices.

crease the volume of sales, to improve production efficiency, and then to increase output to sell at still lower prices (Illustration 1-14). The inventions of Edison (Illustration 1-15) and DeForest of electrical equipment and electron tubes started the widespread use of power systems and communication networks. Following the demonstrations by the Wright brothers that man could build a machine that would fly, aircraft of many types developed rapidly (Illustration 1-16).

These inventions, typical of many basic discoveries that were made early in the century, exemplify the spirit of progress of this period. So fast has been the pace of discovery, with one coming on the heels of another, that it is difficult to evaluate

---

I know this world is ruled by Infinite Intelligence. It required Infinite Intelligence to create it and it requires Infinite Intelligence to keep it on its course. Everything that surrounds us—everything that exists—proves that there are Infinite Laws behind it. There can be no denying this fact. It is mathematical in its precision.
—Thomas Alva Edison

No man really becomes a fool until he stops asking questions.
—Charles P. Steinmetz

Illustration 1-15
Nations the world over acknowledge their indebtedness to Thomas Alva Edison, inventor, and Charles Proteus Steinmetz, electrical engineer, for their significant inventions relating to electricity.

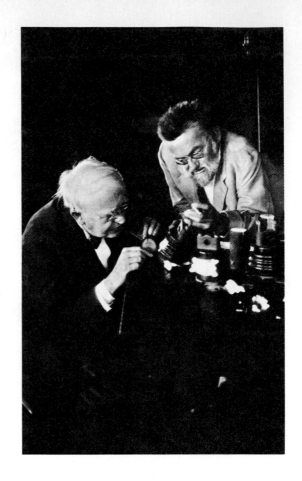

Illustration 1-16
Orville and Wilbur Wright, U.S. inventors and aviation pioneers, achieved in 1903 the world's first successful powered, sustained, and controlled flights of an airplane.

properly their relative importance, although we certainly can realize their impact on our way of life. However, in a number of instances the practicality of an engineering invention has been demonstrated many years in advance of its implementation (Illustration 1-17).

Until late in the nineteenth century, engineering as an applied science was divided into two principal groups, civil and military. Mining and metallurgy was the first group to be recognized as a separate branch, and the American Institute of Mining and Metallurgical Engineers was founded in 1871. In 1880 the American Society of Mechanical Engineers was founded, and in 1884 the American Institute of Electrical Engineers (now the Institute of Electrical and Electronics Engineers) was founded. In 1908 the American Institute of Chemical Engineers was founded and since then a number of other branch societies have been founded with objectives peculiar to specialized fields of engineering endeavor.

An outstanding characteristic of this century is the increased use of power. Albert Einstein was one of the world's most acclaimed physicists. His statement of the equivalence of mass and energy, $E = mc^2$, made possible the emergence of nuclear power (Illustration 1-18). In 1940 it was estimated that the total energy generated in the United States would be the equivalent in "muscle-power energy" of 153 slaves working for every American man, woman, and child in the country. Today a similar calculation would show that about 500 "slaves" are available to serve each person. This is a considerable advance from the days of the Egyptians and Greeks.

One of the most significant inventions of this century is xerography, which has greatly facilitated the communication process for all people. The inventor, Chester F. Carlson, was granted the basic patent in 1940. His machines, and those later pat-

Illustration 1-17
The first public demonstration of television as an adjunct to the telephone took place on April 7, 1927, when Herbert Hoover, then Secretary of Commerce, and other officials in Washington, D.C., spoke "face-to-face" with officials of the Bell Laboratories in New York City. In the ensuing 50 years the concept of a video telephone system has progressed through several evolutionary designs, represented by the models in the foreground.

Illustration 1-18
Albert Einstein, father of the nuclear age, emphasized the importance of simplicity in all matters. He believed that harmony will result if people act in accordance with principles founded on consciously clear thinking and experience.

terned after them, have been especially beneficial to business and industry (Illustration 1-19).

Following World War II, the political, economic, and scientific disorganization in the world caused the emigration of many outstanding educators, scientists, and engineers to the United States. Here they have been able to expand their knowledge

Illustration 1-19
Chester F. Carlson's xerography machine, being demonstrated here, led the way to improved communication for all people everywhere.

and skills and to aid generally in advancing our own understanding of the basic natural laws on which the improved techniques of the future will be based.

Engineering today

Broadly speaking, modern engineering had its beginnings about the time of the close of the Civil War. Within the last century, the pace of discovery has been so rapid that it can be classed as a period within itself. In these modern times, engineering endeavor has changed markedly from procedures used in the time of Imhotep, Galileo, or Ampere. Formerly, engineering discovery and development were accomplished principally by individuals. With the increased store of knowledge available and the widening of the field of engineering to include so many diverse branches, it is usual to find groups or teams of engineers and scientists working on a single project. Where formerly an individual could absorb and understand practically all of the scientific knowledge available, now the amount of information available is so vast that an individual can retain and employ at best only a part of it.

Since 1900 the ratio of engineers and scientists in the United States in comparison to the total population has been steadily increasing. Predictions based upon past increases seem to indicate the following:

Year	Ratio of U.S. engineers and scientists to population
1900	1 to 1800
1950	1 to 190
1960	1 to 130
1980	1 to 65
2000	1 to 35

If this is the case, there will be an even greater increase in technological advance in the next 20 years than there has been in the past 20 years.

Within the past two decades, four technological developments have produced profound changes in our way of life. These developments are nuclear power, the electronic digital computer, interplanetary space navigation, and microelectronics. These concepts are still in their early stages of development, but historians of the future may well refer to our time by such terms as the *nuclear age*, the *computer age,* or the *age of space travel* (Illustration 1-20). The engineer has been a principal developer of these concepts because of the need for their capabilities. The ocean offers great possibilities for technological exploration and perhaps even greater rewards for civilization than has space exploration (Illustration 1-21).

> My interest is in the future because I am going to spend the rest of my life there.
> —Charles F. Kettering

Illustration 1-20
Man's safe exploration of the moon will long stand as one of his greatest engineering achievements.

In the past decade a larger percentage of women have begun to study engineering. Traditionally, particularly in the United States, women have not been so free to select engineering as a career choice. In this mechanized age physical strength is of little importance to the engineer, while mental ability, creativity, curiosity, and interest in problem solving are more important attributes. Both men and women possesss these traits (see Illustration 1-22).

In this age, as in any age, engineers must be creative and must be able to visualize what may lie ahead. They must possess fertile imaginations and a knowledge of what others have done before them. As Sir Isaac Newton is reputed to have said, "If I have seen farther than other men, it is because I have stood on the shoulders of giants." The giants of science and engineering still exist. All any person must do to increase his or her field of vision is to climb up on their shoulders.

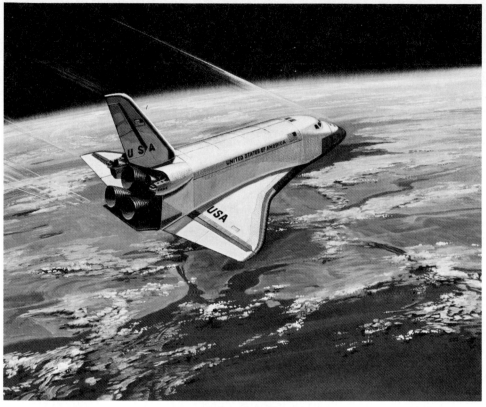

Illustration 1-21
The space shuttle orbiter will carry satellites into space for placement in orbit. It not only has the ability to maneuver in space like a spacecraft, but also to enter the earth's atmosphere and land in a manner similar to a jet transport.

Illustration 1-22
Career opportunities for women in engineering are unlimited.

Problems

1-1. Prepare a chart as a series of columns, showing happenings and their approximate dates in a vertical time scale for various civilizations, beginning about 3000 B.C. and extending to about A.D. 1200.

Chinese	Middle East	Egyptian	Greek	Roman	Western Europe

1-2. Prepare a brief essay on the possible circumstances surrounding the discovery that an iron needle when rubbed on a lodestone and then supported on a bit of wood floating on water will point to the north.

1-3. Determine from historical references the approximate number of years that major civilizations existed as important factors in history.

1-4. What were the principal reasons for the lack of advancement of discovery in Greek science?

1-5. Explain why the development of a successful horsecollar was a major technological advancement.

1-6. Draw to some scale a typical cross section of the "Great Wall" of China, and estimate the volume of rock and dirt required per mile of wall.

1-7. Trace the development of a single letter of our alphabet from its earliest known symbol to the present.

1-8. Describe the details of preparing papyrus from reedlike plants which grew in Egypt.

1-9. Describe the patterns of behavior and accomplishments of ancient engineers that seem to have made successful civilizations, and to have prolonged their existence.

1-10. Prepare lists of prominent persons who contributed outstanding discoveries and developments to civilization during the period from A.D. 1200 to A.D. 1900 in the fields of (*a*) mathematics, (*b*) astronomy, (*c*) electricity, (*d*) mechanics, and (*e*) light.

Person	Date	Major contribution

1-11. List the ten most significant engineering achievements of the twentieth century.

1-12. Beginning with 3000 B.C., list the 25 most significant engineering achievements.

1-13. Based upon your knowledge of world history, describe the probable changes that might have occurred had the airplane not been invented until 1970.

1-14. Describe the precision with which the pyramids of Egypt were constructed. How does this precision compare with that of modern office buildings of more than 50 stories in height?

1-15. Trace the development of the power-producing capability of man from 3000 B.C. to the present.

1-16. Write a 100-word essay about the engineering accomplishments of each of three women engineers.

2

The technological team

Just over a decade ago our civilization was privileged to participate in and help fulfill one of man's oldest desires—to walk on the moon. Indeed it was as engineer–astronaut Neil Armstrong described so aptly, "One small step for man, one giant leap for mankind." For centuries, exploring the surface of the moon had been one of man's cherished dreams. Moreover, the mathematical principles and laws of nature governing whether such a voyage could be possible had been known in scientific circles for many years. Yet it all seemed to be an impossible dream. The lacking ingredient was a broad interdisciplinary technology *of undreamed complexity* that would be capable of fulfilling the requirements of a lunar exploration. And, of course, without a national commitment of purpose to complete such a voyage successfully, the dream would still be unfulfilled.

Today we live in a technological world where such people as housewives, farmers, home builders, accountants, teachers, and physicians carry out their daily tasks in a manner totally different than their counterparts did only 50 years ago. In the case of each worker, one or more *technological teams* have been instrumental in producing new and improved devices, designs, and systems that have eased the task and improved the overall quality of life. These team members are known in our society today as scientists, engineers, technologists, technicians, and craftsmen. In earlier cultures, the attributes of the scientist–engineer–craftsman were most frequently embodied in one person who had the basic knowledge, interest, and skill in the full technological spectrum of activities. Leonardo da Vinci was such a person. Today, partially as a result of the tremendous increase in population that has occurred and somewhat as a result of the overall increased complexity of technology, occupational specialization in a separate career field is more desirable for the majority of people. Thus, it is now more likely that a high school graduate would plan an educational–work-experience program to become a scientist, or an engineer, or a technologist, and

Craftsman	Work involves repetitive and manipulative skills requiring physical dexterity. Rarely supervises the work of others.
Technician	Performs routine equipment checks and maintenance. Carries out plans and and designs of engineers. Sets up scientific experiments. Seldom supervises others.
Technologist	Applies engineering principles for industrial production, construction, and operation. Works with engineering design components. Occasionally supervises others.
Engineer	Identifies and solves problems. An innovator in applying principles of science to produce economically feasible designs. Frequently supervises others.
Scientist	Searches for new knowledge concerning the nature of man and the universe. Infrequently involved in supervisory work.

Figure 2-1 Occupational spectrum.

so on, rather than some combination. A purpose of this chapter is to further clarify the types of work that each member of the technological team will likely perform, and to list some of the personality traits and aptitudes that are indicative of success for each career field. Figure 2-1 depicts graphically this occupational spectrum.

We must recognize that as individuals we do not always fit neatly into one specific area of the technological occupational spectrum. Rather, our interests and aptitudes may bridge across one or two of the work areas—or quite naturally across the entire spectrum. If so, good! However, educational programs leading to the various career choices are usually quite different in their composition. In most schools they strengthen the interests and preferences depicted in Figure 2-2.

In a general sense Figure 2-2 indicates how such factors as degree of aptitude for manipulating mathematical expressions and the degree of satisfaction that is derived from manual artistry will influence the probability of lasting interest in these various career fields.

During the last 100 years, by whatever names they may have been called, members of the technological team have been largely responsible for the development in this country of the highest standard of living that the world has ever known. This cooperative effort has made possible the solution of practical problems on a scale never accomplished before in all recorded history. Each member of the team has been

Figure 2-2

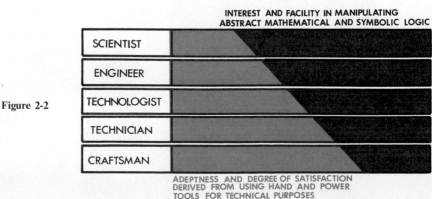

important in this effort. However, the craftsman was the first to appear on the American scene.

The craftsman

In colonial times the master craftsman held a position of considerable esteem in the community. Usually, he operated his own shop, determined the designs, set standards, and directed the apprentice workmen and other artisans who worked for him. The period of apprenticeship was usually about seven years, and individual skills were mastered by constant repetition. Paul Revere, an outstanding master craftsman of the colonial era, was skilled in the casting and working of metal, particularly silver and copper.

Since the products produced were usually sold, the average craftsman was sensitive to the style, marketability, and serviceability required by his customers. He was, therefore, accustomed to working within definite limits. Examples of craftsmen of this period were silversmiths, weavers, cabinetmakers, potters, blacksmiths, and candlestick makers. They were the backbone of early American industry.

As American industry became more industrialized, the need for master craftsmen skills diminished and the role of the craftsman changed. Machines were developed to do many of the routine jobs, and individual workers found themselves to be more captives of the machines than the reverse. Also, as factories grew a worker no longer was responsible for making a product from start to finish. Now he specialized in a

Illustration 2-1
Welding craftsmen are important contributors to the successful completion of many engineering designs.

single task—one of a series of operations—and he performed this task over and over to achieve the desired production. When this happened, the pride of accomplishment that is so necessary to motivate the craftsman disappeared. Because of this trend many individuals who aspired to be true craftsmen moved into other areas of work.

Today craftsmen continue to work in a variety of technologically related areas. These do provide the same intense feeling of pride and satisfaction for the individual as did the domestic-type craft work of several decades ago. Among these are carpenters, welders, tool and die makers, pattern makers, and precision machinists. A high school education is needed for entry into these specialty areas. In most cases this is followed by an intensified period of skill training that may vary from 6 to 18 months.

The technician

The engineering technician is interested in working with equipment and in assembling the component parts of designs that are designed by others. In this respect the technician is an experimentalist—an Edison-type thinker rather than an Einstein- or Steinmetz-type thinker. Preference is given to assemble, repair, or to the making of improvements in technical equipment by learning its characteristics rather than by studying the scientific or engineering basis for its original design.

A technician possesses many of the skills of a craftsman, and frequently is personally able to effect required physical changes in engineering hardware. Because most work is concerned directly with such equipment, it is often possible to suggest answers to difficult problems that have not been thought of previously.

The technician will probably work in a laboratory, out in the field on a construction job, or perhaps troubleshoot on a production line, rather than, for example, work at a desk. Of course, some technicians do their work in drafting rooms or offices under the direction of engineers. They also are frequently found in research laboratories, where they render effective service in repairing equipment, setting up experiments, and accumulating scientific data. They are very important in this role because it is often the case that scientists themselves are not too adept or interested in carrying out tasks of this type.

In construction work, technicians also play key roles. They are needed to accomplish tasks in surveying, to make estimates of material and labor costs, to be responsible occasionally for the coordination of skilled labor and the work of subcontractors, and for the delivery of materials to the job sites.

A substantial number of engineering technicians are employed in the manufacturing and electronics industry and by public utilities. They may carry out standard calculations, serve as technical salesmen, make estimates of costs, or assist in preparing service manuals, such as for electronic equipment and plant operation and maintenance. They install and maintain, make checks on, and frequently modify electrical and mechanical equipment. As a group they are important problem-solving-oriented persons whose interests are directed more to the practical than to the theoretical. Technicians learn fundamental scientific theory and master the mathematical topics most useful in analyzing and solving problems. The majority of their studies, however, are more practically oriented.

The technician's education typically requires two full years of collegiate-level study. Generally, this work is taken in a technical institute or community college and leads to an associate degree in technology. In many instances this school work is transferable to a senior academic institution and may be applied as credit toward a bachelor of engineering technology program. In most cases, however, these courses would not be applicable toward a degree in engineering or science.

Illustration 2-2
Technicians perform tasks that may be both routine and complex. Here the technician is making final adjustments on a geostationary meteorological satellite used to track changing weather conditions.

The technologist

The engineering technologist is the most recent of the technological team to appear, having emerged in the last two decades. Engineering technology is that part of the technological spectrum which requires the application of scientific and engineering knowledge and methods combined with technical skills in support of engineering activities. Technologists work in the occupational domain between the craftsman-technician and the engineer. Their areas of interest typically are less theoretical and mathematically based than those of the engineer. Some of these are construction, operation, maintenance, and production. Also, as a group they are more inclined to be known as design organizer–producers, rather than design innovators.

Ordinarily the engineering technologist will work more with the development of various design components of systems that have been designed and developed by engineers. They are valuable in research for liaison work. They are also well suited to assume a variety of technical supervisory and management roles in industry.

Engineering technologists are graduates of baccalaureate degree programs in technology. The programs exist in two forms. One is a separate and distinctive

Illustration 2-3
Technologists must have a fundamental understanding of the theoretical principles involved as well as a facility for working with the associated engineering equipment.

four-year curriculum that emphasizes the solution of practical engineering problems. The other has similar objectives but is a two-year extension of the two-year associate degree program for engineering technicians.

The engineer

The engineer is a person who enjoys changing the status quo to gain an improvement. Two functions are performed in this role: first, recognition, identification, and definition of problems and deficiencies, or conditions that need improvement; and second, using one's reservoir of knowledge and skill in innovative thought to produce one or more acceptable solutions.

Thus, as a personal characteristic the engineer should possess a persistently inquiring and creative mind—yet one that is receptive to producing realistic innovative action. This must include a fundamental understanding of the laws of nature and of mathematics, as well as being able to recognize and interpret their application in real situations. Although these characteristics are most often associated with those who provide our nation technological leadership in commerce and industrial production, they are also highly desired qualities to attain for those who want to follow careers in management, medicine, and law.

The baccalaureate degree in engineering is required for entry into industry, and the masters degree (a total of five years of college) is preferred for many types of work.

> Thousands of engineers can design bridges, calculate strains and stresses, and draw up specifications for machines, but the great engineer is the man who can tell *whether* the bridge or the machine should be built at all, *where* it should be built, and *when*.
> —Eugene G. Grace

Illustration 2-4
Engineers create custom, special-purpose designs that have never existed previously, like the lunar vehicle pictured here.

Illustration 2-5
The primary objective of the scientist is to discover new knowledge. Here a research scientist is working on an experiment to test the feasibility of processing glass in space.

> Scientists study the world as it is, engineers create the world that never has been.
> —Theodore von Karman

The scientist

The primary objective of the scientist is to discover, to expand existing fields of knowledge, to correlate observations and experimental data into a formulation of laws, to learn new theories and to explore their meanings, and in general to broaden the horizons of science. The scientist is typically a theoretician who is concerned about *why* natural phenomena occur. As suggested above, the scientist's primary objective is to expand the world's reservoir of knowledge, and there is little reason to pause to search for practical uses for any new-found truths that may be discovered. There are, of course, *applied scientists* who work to find specific uses for new knowledge, such as devising a new instrument or synthesizing an improved medicine. For this reason it is difficult for the average person to tell the difference between an applied scientist and an engineer. In terms of schooling, the person educated as a scientist will generally have earned a doctorate (seven years of college minimum) in some field of science.

The roles of the engineer and technologist

It is generally known that technological advances have made possible the many material things that now make our lives more enjoyable and provide extra time for recreation and study. It is also an accepted belief that engineers and technologists will continue to provide innovative and creative designs for the purpose of easing the burden of man's physical toil and to convert the materials and forces of nature to the use of all mankind; however, when the time comes to make a choice of careers, each student finds himself groping for an answer to the question, "As an engineer, technologist, or technician, how would *I* fit into this picture."

First, we must realize that one does not become a graduate solely by studying a few courses. A technical education is more than knowing when to manipulate a set of formulas, where to search in an armful of reference books, or how to get accurate answers from a calculator or computer; it is also a state of mind. For example, through experience and training the engineer must be able to formulate problem statements and must conceive design solutions that many times involve novel ideas and creative thought processes. The engineer must also exercise judgment and restraint, design with initiative and reliability, and be completely honest individually and with others. These qualities should all mature as the student engineer advances from elementary to graduate-level studies.

What skills do engineers and technologists need?

The public expects engineers and technologists to be competent technically. Through the years the profession has built up a record of producing things that work. No one expects a company to produce television tubes that explode spontaneously, or a bridge that falls down, or an irrigation ditch that has the wrong slope. In fact, major technological failures are so rare that when they do occur they usually are front-page news.

The college engineering and technology curricula are designed to instill technical competence. The grading system generally used rewards acceptable solutions and penalizes inferior or unworkable solutions to practical problems. The subjects studied are not easy to master, and usually those persons who do not adapt themselves to the discipline of study and who do not accept the ideas of the exactness of Nature's laws will not complete a college program in engineering or technology. Such discipline has paid handsome dividends. The consistency of quality in graduates over many years has given the public a confidence that results in large measure from the rigorous but realistic accreditation standards of the Accreditation Board for Engineering and Technology (ABET).

It should also be realized, however, that the completion of a baccalaureate degree must not be the end of study. The pace of discovery is so rapid today that, even with constant study, one can barely keep abreast of technological improvements. If the graduate should not make an effort to keep abreast of technological developments, deterioration of currency will likely be apparent within five years.

Engineers and technologists, then, must be capable of dealing with technological problems—not only those which they may have been trained to handle in college, but also new and unfamiliar situations that arise as a result of new discoveries. Of course, in preparation for the solution of real problems college courses can present problem areas only in general terms. Graduates are expected to provide solutions to new problem situations using a base of fundamental principles and an understanding of the most effective methods to use in problem solving.

Illustration 2-6
Many of today's complex engineering problems could not be solved without the team effort of technicians, technologists, and engineers.

What are the opportunities after college?

A question that arises frequently in the mind of high school graduates is, "What if I begin in engineering and decide later to change to another course of study? What would be the consequences?" Let us examine some of the possibilities. Normally an engineering student will follow engineering as a profession. However, many students change their mind during their college career or after graduation. Many authorities agree that engineering courses are excellent training for a great variety of careers, and records reveal that perhaps as many as 40 per cent of people on a management level have engineering educations. One of the basic and most valuable training concepts of an engineering education is that *engineering students are taught to think logically*. This means that, if a later career decision is made not to following engineering as a profession, the training and experience gained in studying engineering courses still will prepare a person for a wide variety of occupations.

What will the future hold for the serious student who chooses engineering as a career? *First,* employment possibilities will be good. The rigor of the college courses usually removes those who are unable or unwilling to stick with a problem until they come up with a reasonable answer. Those who graduate in engineering usually are well qualified technically and, in addition, are well rounded in their knowledge of nonengineering courses.

Second, the work experience will be satisfying. A sampling of questionnaires sent to engineers in large industrial concerns shows that those with several years experience almost unanimously enjoy their work. They like the opportunities for advancement, the challenges of new and exciting problems to be solved, the friendships gained in contacts with people with diverse backgrounds, and the possibilities of seeing their ideas develop into working realities. The engineer will find that, as a profession, the salary scale is among the higher groups, and individual income usually is determined largely by the quality of one's own efforts.

Third, engineering provides unlimited opportunities for creative design. As has been mentioned before, the pace of discovery is so great that the need for the application of discoveries provides countless places where the engineer with intitiative and creative ability can spend time in idea development. Also, in applications which are old and well known, the clever engineer can devise new, better, and more economical ways of providing the same services (Illustration 2-7). For example, although roads have been built for centuries, the need for faster and more efficient methods of road building offers a continuing challenge to engineers.

Sometimes nontechnical people say that engineers are too dogmatic, that they think of things as being either positive or negative. To a certain extent this is true because of the engineer's training. Engineers are educated to give realistic answers to real problems and to make the answers the most practical ones that can be produced. Within an acquired knowledge of nature's laws, the engineer should be able to obtain a precise solution to a given problem, and be willing to defend that solution. To a nontechnical person accustomed to arriving at a solution by surmise, argument, and compromise, the positive approach of the engineer frequently is distressing. Part of the postgraduate training of an engineer is learning to convince nontechnical people of the worth of a design. The ability to reason, to explain by using simple applications, and to have patience in presenting ideas in simple terms that can be understood are essential qualities of the successful engineer.

As an engineer you are not expected to be a genius. However, by training, you are a leader. You have a responsibility to your profession and to your community to exercise that leadership, and you should establish a set of technical and moral stand-

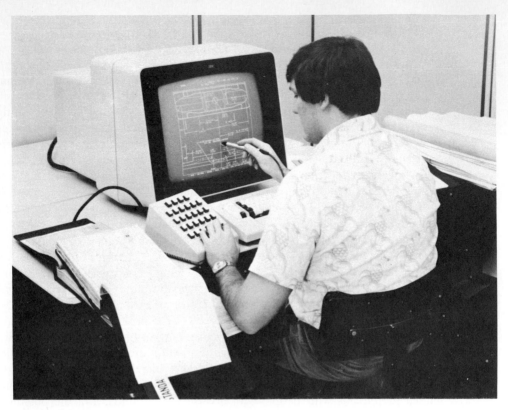

Illustration 2-7
Today engineers use computer aided design (CAD) equipment and procedures to accelerate and greatly improve their design capability.

ards that will provide a wholesome influence upon all levels of your organization and upon your community as a whole.

Problems

2-1. Interview an engineer and write a 500-word essay concerning his or her work.

2-2. Survey the job opportunities for engineers, scientists, and technicians. Discuss the differences in opportunity and salary.

2-3. Discuss the role of the engineer in government.

2-4. Frequently, technical personnel in industry are given the title "engineer" in lieu of other benefits. Discuss the difficulties that arise as a result of this practice.

2-5. Write an essay on the differences between the work of the engineer, the scientist, and the technician.

2-6. Write an essay on the differences between the education of the engineer, the scientist, the technologist, and the technician.

2-7. Interview an engineering technician and write a 500-word essay concerning her work.

2-8. Discuss the role of the engineering technician in the aircraft industry.

2-9. Investigate the opportunities for employment of electronic technicians. Write a 500-word essay concerning your findings.

2-10. Investigate the differences in educational requirements of the engineer and the technician. Discuss your conclusions.

2-11. Classify the following items as to the most probable assignment in the occupational spectrum:

    a. Detail drawing of a small metal mounting bracket.

    b. Assisting an engineer in determining the pH of a solution.

    c. Boring a hole in an aluminum casting to fit a close-tolerance pin.

    d. Determining the behavior of flow of a viscous fluid through a pipe elbow.

    e. Determining the percentage of carbon in a series of steel specimens to be used in fabricating cutter bits.

    f. Designing a device to permit the measurement of the temperature of molten zinc at a location approximately 150 feet from a vat.

    g. Preparation of a laboratory report concerning the results of a series of tests on an assortment of prospective heat-curing bonding adhesives.

    h. Preparing a work schedule for assigning personnel for a 2-day test of a small gasoline engine.

    i. Determining the effects of adding ammonia to the intake air of a gas turbine engine.

    j. Fabrication of 26 identical transistorized circuits using printed circuit boards.

    k. Calculation of the area of an irregular tract of land from a surveyor's field notes.

    l. Preparation of a proposal to study the effect of sunlight on anodized and unanodized aluminum surfaces.

    m. Design and fabrication of a device to indicate the rate of rainfall at a location several hundred feet from the sensing apparatus.

    n. Interpretation of the results of a test on a punch-tape-controlled milling machine.

    o. Preparing a computer program to determine the location of the center of gravity of an airplane from measured weight data.

2-12. A large office building is to be constructed, and tests of the load-bearing capacity of the underlying soil are to be made. Outline at least one way each member of the scientific team (scientists, engineers, technologists, technicians, and craftsmen) is involved in the work of determining suitability of the soil for supporting a building.

3

Career fields in engineering and technology

Much of the change in our civilization in the past 100 years has been due to the work of the engineer. We hardly appreciate the changes that have occurred in our environment unless we attempt to picture the world of a few generations ago, without automobiles, telephones, radios, electronics, transportation systems, supersonic aircraft, automatic machine tools, electric lights, television, and all the modern appliances in our homes. In the growth of all these things the role of the engineer is obvious.

Development in the field of science and engineering is progressing so rapidly at present that within the last 10 years we have acquired materials and devices that are now considered commonplace but which were unknown to our parents. Through research, development, and mass production, directed by engineers, ideas are made into realities in an amazingly short time.

The engineer is concerned with more than research, development, design, construction, and the operation of technical industries, however, since many are engaged in businesses that are not concerned primarily with production. Formerly, executive positions were held almost exclusively by persons whose primary training was in the field of law or business, but the tendency now is to utilize engineers more and more as administrators and executives.

No matter what kind of work the engineer may wish to do, there will be opportunities for employment not only in purely technical fields but also in other functions, such as general business, budgeting, rate analysis, purchasing, marketing, personnel, labor relations, and industrial management. Other opportunities also exist in such specialized fields of work as teaching, writing, patent practice, and work with the military establishment.

Although college engineering curricula contain many basic courses, there will be some specialized courses available that are either peculiar to a certain curriculum or are electives. These specializations permit each student to acquire a particular proficiency in certain technical subjects so that, for example, professional identification can be acquired in electrical, civil, chemical, mechanical, or industrial engineering.

Education in the application of certain subject matter to solve technological problems in a certain engineering field constitutes engineering specialization. Such training is not for manual skills as in trade schools, but rather is planned to provide preparation for research, design, operation, management, testing, maintenance of projects, and other engineering functions in any given specialty.

The principal engineering fields of specialization that are listed in college curricula and that are recognized in the engineering profession are described in the following sections.

Aerospace and astronautical engineering

Powered flight began in 1903 at Kitty Hawk, North Carolina. Perhaps no other single technological achievement has been so significant for mankind. Through faster transportation and improved communications almost every aspect of our daily life has been affected. However, not all challenges are associated with spaceflight. Problems associated with conventional aircraft, and the development of special vehicles such as hydrofoil ships, ground-effect machines, and deep-diving vessels for oceanographic research are all concerns of the industry.

Within the past few years many changes have taken place which have altered the work of the aeronautical engineer—not the least of which is man's successful conquest of space. Principal types of work vary from the design of guided missiles and spacecraft to analyses of aerodynamic studies dealing with the performance, stability, control, and design of various types of planes and other devices that fly (Illustration 3-1). Most of such activity is concerned with the design, development, and performance testing of supersonic commercial transports and their propulsion systems.

Although aerospace engineering is one of the newer fields, it offers many possibilities for employment. Continued exploration and research in previously uncharted areas is needed in the fields of propulsion, materials, thermodynamics, cryogenics, navigation, cosmic radiation, and magnetohydrodynamics. It is predicted that within the near future the chemically fueled rocket engine, which has enabled man to explore lunar landscapes, will become obsolete as the need increases to cover greater and greater distances over extended periods of time.

The rapidly expanding network of airlines, both national and international, provides many openings for the engineering graduate. Since the demand for increasing numbers of aircraft of various types exists, there are opportunities for work in manufacturing plants and assembly plants and in the design, testing, and maintenance of aircraft and their component parts. The development of new types of aircraft, both civilian and military, requires the efforts of well-trained aeronautical engineers, and it is in this field that the majority of positions exists. Employment opportunities exist for specialists in the design and development of fuel systems using liquid oxygen propellants and solid propellants. Control of the newer fuels involves

Illustration 3-1
Aeronautical engineers frequently are called upon to design unconventional flying machines.

precision valving and flow sensing at very low and very high temperatures. Air traffic control is a problem that is becoming increasingly more complex, and trained people are needed here. The design of ground and airborne systems that will permit operation of aircraft under all kinds of weather conditions is also a part of the work of aeronautical engineers.

The aerospace engineer works on designs that are not only challenging and adventuresome but also play a major role in determining the course of present and future world events.

Agricultural engineering

Agricultural engineering is that discipline of engineering that spans the area between two fields of applied science—agriculture and engineering. It is directly concerned with supplying the means whereby food and fiber are supplied in sufficient quantity to fill the basic needs of all mankind. In the next 30 years the world's population is expected to double. This factor, plus the increasing demands of people throughout the world for increased standards of living, provides unparalleled challenges to the

Figure 3-1

"... All I can say Capcom is that these things weren't there when we put it out last night ..."

agricultural engineer. Not only must the quantity of food and fiber be increased, but the efficiency of production also must be steadily improved in order that personnel may be released for other creative pursuits. Through applications of engineering principles, materials, energy, and machines may be used to multiply the effectiveness of man's effort. This is the agricultural engineer's domain.

Illustration 3-2
In this picture agricultural engineers at a research center test the safety features of a farm tractor. Agricultural engineers apply fundamental engineering principles of analysis and design to improve our methods of food production and land utilization.

Figure 3-2

"Hmph! ... I've told you Eli, you can't make gin from cotton!"

In order that the agricultural engineer may understand the problems of agriculture and the application of engineering methods and principles to their solution, instruction is given in agricultural subjects and the biological sciences as well as in basic engineering. Agricultural research laboratories are maintained at schools for research and instruction using various types of farm equipment for study and testing. The young person who has an analytical mind and a willingness to work, together with an interest in the engineering aspects of agriculture, will find the course in agricultural engineering an interesting preparation for his or her life's work.

Many agricultural engineers are employed by companies that serve agriculture and some are employed by firms that serve other industries. Opportunities are particularly apparent in such areas as (1) research, design, development, and sale of mechanized farm equipment and machinery; (2) application of irrigation, drainage, erosion control, and land and water management practices; (3) application and use of electrical energy for agricultural production, and feed and crop processing, handling, and grading; (4) research, design, sale, and construction of specialized structures for farm use; and (5) the processing and handling of food products.

Architectural engineering

The architectural engineer is interested primarily in the selection, analysis, design, and assembly of modern building materials into structures that are safe, efficient, economical, and attractive. The education received in college is designed to teach

Illustration 3-3
Architectural engineers must be equally cognizant of aesthetic and structural design considerations.

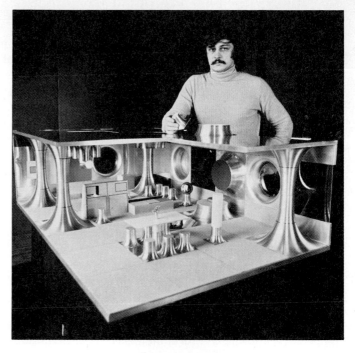

Figure 3-3

"I know you're new to architecture Wick, but surely they taught you how the draw in "drawbridge" is supposed to work ..."

one how best to use modern structural materials in the construction of tall buildings, manufacturing plants, and public buildings.

The architectural engineer is trained in the sound principles of engineering and at the same time is given a background supportive of the point of view of the architect. The architect is concerned with the space arrangements, proportions, and appearance of a building, whereas the architectural engineer is more nearly a structural engineer and is concerned with safety, economy, and sound construction methods.

Opportunities for employment will be found in established architectural firms, in consulting engineering offices, in aircraft companies, and in organizations specializing in building design and construction. Excellent opportunities await the graduate who may be able to associate with a contracting firm or who may form a partnership with an architectural designer. In the field of sales an interesting and profitable career is open to the individual who is able to present ideas clearly and convincingly.

Bioengineering

Bioengineering encompasses all aspects of the application of engineering methods to the use and control of biological systems. It bridges the engineering, physical, and life sciences in identifying and solving medical and health-related problems. Bioengineers

Illustration 3-4
The work of bioengineers has made possible the development of many life-lengthening and life-enhancement systems.

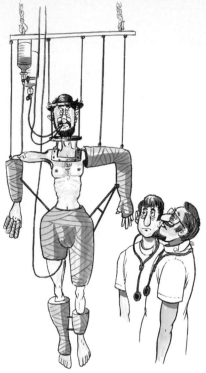

Figure 3-4

"Poor devil broke every bone in his body . . . fortunately we had a resident bioengineer . . ."

are team players in much the same way as many athletes. For example, engineers, physicists, chemists, and mathematicians routinely join with the biologist and physician in developing techniques, equipment, and materials.

The range of the bioengineers' interests is very broad. It would involve, for example, the development of highly specialized medical instruments and devices—including artificial hearts and kidneys and the development of lasers for surgery and cardiac pacemakers that regulate the heartbeat. Other biomedical engineers may specialize more particularly in the adaptation of computers to medical science, such as in monitoring patients or processing electrocardiograph data. Some will design and build systems to modernize laboratory, hospital, and clinical procedures.

Those selecting a career in bioengineering should anticipate earning a graduate degree since advanced study beyond the bachelor's degree is acutely needed to attain a depth of knowledge from at least two diverse disciplines.

At present bioengineering is a small field because few engineers have attained the necessary depth of academic training and experience in the life sciences. Therefore, job opportunities for graduates are excellent. Here indeed is a promising new field for those so inclined.

Ceramic engineering

Today our technological world is amazingly dependent upon ceramics of all types. Unlike many other products they appear in every part of the spectrum of life, from

Figure 3-5

"When our competition went to ceramic insulating tiles for spacecraft, we moved to the southwest and started our own 'coil' method ..."

beautiful but commonplace table settings, to the protective coatings of electrical transducers or the refractories of space exploratory rocket nozzles, to the spark plugs of a farmer's tractor. Exactly what are ceramics? When did man first find a use for them?

Ceramics are nonmetallic, inorganic materials that require the use of high temperatures in their processing. In the earliest form, clay pottery of 10,000 B.C. has been found to be excellently preserved. The most common of ceramics, glass—an ancient discovery of the Phoenicians (about 4000 B.C.), is a miracle material in every sense. It may be made transparent, translucent, or opaque, weak and brittle or flexible and stronger than steel, hard or soft, water soluble or chemically inert. Truly it is one of the most versatile of engineering materials.

Although it is imperative today that all engineers have a fundamental understanding of the adaptive use of ceramics, it is ceramic engineers who are expert in the development and production of ceramic materials. Their activities cover a wide range of activities from the conception of the initial idea to the development, production, evaluation, application, and sale of the product. Therefore, ceramic engineers are employed by a variety of industries, from the specialized raw material and ceramic product manufacturers to the chemical, electrical and electronic, automotive, nuclear, and aerospace industries.

Illustration 3-5

Ceramic engineers hold in their hands a very important answer to the impending crisis in the shortage of metallic materials.

Chemical engineering

Chemical engineering is responsible for new and improved products and processes that affect every person. This includes materials that will resist extremities of heat and

Illustration 3-6

Chemical engineers apply the results of chemical research in speeding the identification of air and water pollutants and in eliminating chemical contaminants in food and industrial products.

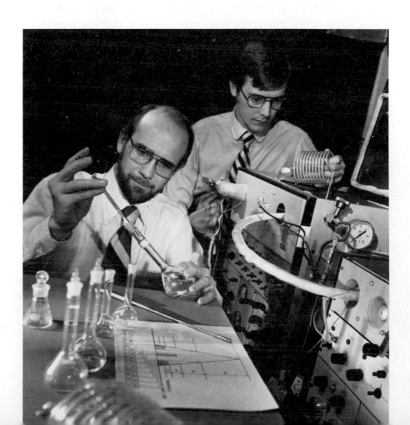

Figure 3-6

*"... I know I ain't a chemical engineer like you son, but
just remember that my design put you through school ..."*

cold, processes for life-support systems in other environments, new fuels for reactors, rockets, and booster propulsion, medicines, vaccines, serum, and plasma for mass distribution, and plastics and textiles to serve a multiplicity of human needs. Consequently, chemical engineers must be able to apply scientifically the principles of chemistry, physics, and engineering to the design and operation of plants for the production of materials that undergo chemical changes during their processing.

The courses in chemical engineering cover inorganic, analytical, physical, and organic chemistry in addition to the basic engineering subjects; and the work in the various courses is designed to be of a distinctly professional nature and to develop capacity for original thought. The industrial development of our country makes large demands on the chemical engineer. The increasing uses for plastics, synthetics, and building materials require that a chemical engineer be employed in the development and manufacture of these products. While well trained in chemistry, the chemical engineer is more than a chemist in that he or she applies the results of chemical research and discovery to the use of mankind by adapting laboratory processes to full-scale manufacturing plants.

The chemical engineer is instrumental in the development of the newer fuels for turbine and rocket engines. Test and evaluation of such fuels and means of achieving production of suitable fuels are part of the work of a chemical engineer. This testing must be carefully controlled to evaluate the performance of engines before the fuel is considered suitable to place on the market.

Opportunities for chemical engineers exist in a wide variety of fields of manufacture. Not only are they in demand in strictly chemical fields but also in nearly all types of manufacturing. The production of synthetic rubber, the uses of petroleum products, the recovery of useful materials from what was formerly considered waste products, and the better utilization of farm products are only a few of the tasks that will provide work for the chemical engineer. Although the first professional work of a chemical engineering graduate may be in production, other opportunities exist in the fields of engineering design, research and development, patents, and sales engineering.

Civil engineering

The civil engineer plans, designs, constructs, and operates physical works and facilities that are deemed essential to modern life. These include the broad categories of construction, soil mechanics and foundations, transportation systems, water resources, sanitation, city planning and municipal engineering, and surveying and mapping. (Illustration 3-7). Construction engineering is concerned with the design and supervision of construction of buildings, bridges, tunnels, and dams. The construction industry is America's largest industry today. Soil mechanics and foundation investigations are essential not only in civilized areas but also for successful conquest of

Illustration 3-7
The civil engineer pictured here is marking a structural member which will be used in the erection of a building that she has designed.

Figure 3-7

*"... The solution to the freeway interchange problem came
to me last night in an Italian restaurant ..."*

new lands such as Antarctica and the lunar surface. Transportation systems include the planning, design, and construction of necessary roads, streets, thoroughfares, and superhighways. Engineering studies in water resources are concerned with the improvement of water availability, harbor and river development, flood control, irrigation, and drainage. Pollution is an ever-increasing problem, particularly in urban areas. The sanitary engineer is concerned with the design and construction of water supply systems, sewerage systems, and systems for the reclamation and disposal of wastes. City planning and municipal engineers are concerned primarily with the planning of urban centers for the orderly, comfortable, and healthy growth and development of business and residential areas. Surveying and mapping are concerned with the measurements of distances over a surface (such as the earth or the moon) and the location of structures, rights-of-way, and property boundaries.

Civil engineers engage in technical, administrative, or commercial work with manufacturing companies, construction companies, transportation companies, and power companies. Other opportunities for employment exist in consulting engineering offices, in city and state engineering departments, and in the various bureaus of the federal government.

Electrical engineering

Electrical engineering is concerned, in general terms, with the utilization of electric energy. It is divided into broad fields, such as information systems, automatic control, and systems and devices. Electricity used in one form or another reaches nearly all our daily lives and is truly the servant of mankind.

The electrical engineer applies sound engineering principles, both mechanical and electrical, in the design and construction of computers and auxiliary equipment. The basic requirements of a computer constantly change and new designs must provide for these necessary capabilities. In addition, a computing machine must be built that will furnish solutions of greater and greater problem complexity and at the same time have a means of introducing the problem into the machine in as simple a manner as possible.

There are many companies that build elaborate computing machines, and employment possibilities in the design and construction part of the industry are not limited. Many industrial firms, colleges, and governmental branches have set up computers as part of their capital equipment, and opportunities exist for employment as computer applications engineers, who serve as liaison between computer programmers and engineers who wish their problems evaluated on the machines. Of course, in a field expanding as rapidly as computer design, increasing numbers of employment opportunities become available. More and more dependence will be placed on the use of computers in the future, and an engineer specializing in this work will find ample opportunity for advancement.

The automatic control of machines and devices, such as autopilots for spacecraft and missiles, has become a commonplace requirement in today's technically con-

Illustration 3-8
Just a few years ago, transistors (each about one centimeter in length) were acclaimed as being responsible for revolutionizing the electronics industry. Today electrical engineers have miniaturized circuitry so that over 27,000 transistors can be placed on a single solid-state chip, as pictured here.

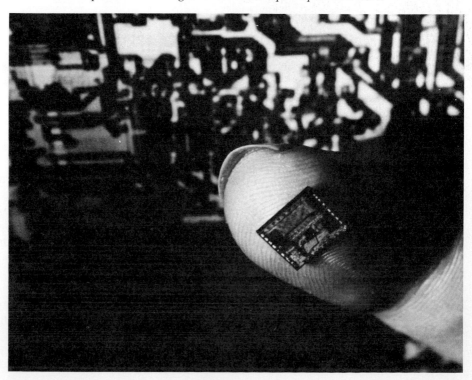

Figure 3-8

As a discipline, electrical engineering has many dissimilar components. However, as with the body parts of the elephant, they each contribute to make up the nation's largest engineering group.

scious society. Automatic controlling of machine tools is an important part of modern machine shop operation. Tape systems are used to furnish signals to serve units on automatic lathes, milling machines, boring machines, and other types of machine tools so that they can be programmed to perform repeated operations. Not only can individual machines be controlled but also entire power plants can be operated on a program system. The design of these systems is performed usually by an electrical or mechanical engineer.

Energy conversion systems, where energy is converted from one form to another, also are a necessity in almost every walk of life. Power plants are constituted to convert heat energy from fuels into electrical energy for transmission to industry and homes. In addition to power systems, communication systems are a responsibility of the electrical engineer. Particularly in communications the application of modern electronics has been most evident. The electrical engineer who specializes in electronics will find that the majority of communication devices employs electronic circuits and components.

Other branches of electrical engineering that may include power or communication activities, or both, are illumination engineering, which deals with lighting using electric power; electronics, which has applications in both power and communications; and such diverse fields as x-ray, acoustics, and seismograph work.

Employment opportunities in electrical engineering are extremely varied. Electrical manufacturing companies use large numbers of engineers for design, testing, research, and sales. Electrical power companies and public utility companies require a staff of qualified electrical engineers, as do the companies which control the networks of telegraph and telephone lines and the radio systems. Other opportunities for employment exist with oil companies, railroads, food processing plants, lumbering enterprises, biological laboratories, chemical plants, and colleges and universities. The aircraft and missile industries use engineers who are familiar with circuit design and employment of flight data computers, servomechanisms, analog computers, vacuum tubes, transistors, and other solid-state devices. There is scarcely any industry of any size that does not employ one or more electrical engineers as members of its engineering staff.

Industrial engineering

Industrial engineers determine the most effective methods of using personnel, machines, and materials in a production environment. Whereas other branches of engineering tend to specialize in some particular phase of science, the realm of industrial engineering may include parts of all engineering fields. The industrial

Illustration 3-9
"I agree—it works. But at that price no customer will ever buy it! We've got to redesign and cut costs."

engineer then will be more concerned with the larger picture of management of industries and production of goods than with the detailed development of processes.

The work of the industrial engineer is rather wide in scope. Generally, work is with people and machines, and because of this it is important that one be educated in both personnel administration and in the relations of people and machines to production.

The advent of the electronic digital computer and other electronic support equipment has revolutionized the business world. Many of the resultant changes have been made as a result of industrial engineering designs. Systems analysis, operations research, statistics, queuing theory, information theory, symbolic logic, and linear programming are all mathematics-based disciplines that are used in industrial engineering work.

The industrial engineer must be capable of preparing plans for the arrangement of plants for best operation and then of organizing the workers so that their efforts will be coordinated to give a smoothly functioning unit. In such things as production lines, the various processes involved must be timed perfectly to ensure smooth operation and efficient use of the worker's efforts. In addition to coordination and automating of manufacturing activities, the industrial engineer is concerned with the development of data processing procedures and the use of computers to control

Figure 3-9

*"As the industrial engineer on this job, I must say your construction methods are the **most** innovative I've seen."*

production, the development of improved methods of handling materials, the design of plant facilities and statistical procedures to control quality, the use of mathematical models to stimulate production lines, and the measurement and improvement of work methods to reduce costs.

Opportunities for employment exist in almost every industrial plant and in many businesses not concerned directly with manufacturing or processing goods. In many cases the industrial engineer may be employed by department stores, insurance companies, consulting companies, and as engineers in cities. The industrial engineer is trained in fundamental engineering principles, and as a result may also be employed in positions which would fall in the realm of the civil, electrical, or mechanical engineer.

The courses prescribed for the student of industrial engineering follow the pattern of the other branches of engineering by starting with a thorough foundation in the engineering sciences. The engineering courses in the later semesters will be of a more general nature, and the curriculum will include such courses as economics, psychology, business law, personnel problems, and accounting principles.

Mechanical engineering

Mechanical engineering deals with power and the design of machines and processes used to generate power and apply it to useful purposes. These designs may be simple or complex, inexpensive or expensive, luxuries or essentials. Such items as the kitchen

Illustration 3-10
Mechanical engineers work with a wide range of mechanical and electrical equipment. Here, new amorphous metal alloys are being developed for use in motors and transformers.

food mixer, the automobile, air-conditioning systems, nuclear power plants, and interplanetary space vehicles would not be available for human use today were it not for the mechanical engineer. In general, the mechanical engineer works with systems, subsystems, and components that have motion. The range of work that may be classed as mechanical engineering is wider than that in any of the other branches of engineering, but it may be grouped generally under two heads: work that is concerned with power-generating machines, and work that deals with machines that transform or consume this power in accomplishing their particular tasks. The utilization of solar energy for domestic and industrial uses is one of the more important areas of the mechanical engineer's work at this time.

There are several general subdivisions of mechanical engineering. Power or combustion engineers deal with the production of power from fuels. Design specialists may work with parts that vary in size from the microscopic part of the most delicate instrument to the massive part of heavy machinery. Included are the mass transit systems that are rapidly becoming a part of our nationwide transportation system. Automotive engineers work constantly to improve the vehicles and engines that we now have. Heating, ventilating, air-conditioning, and refrigeration engineers deal with the design of suitable systems for making our buildings more comfortable and for providing proper conditions in industry for good working conditions and efficient machine operation.

Employment may be secured by mechanical engineering graduates in almost every type of industry. Manufacturing plants, power-generating stations, public utility companies, transportation companies, airlines, and factories, to mention only a

Figure 3-10

A NON-ELECTRIC, AUTOMATIC BROOM.
METRONOME WITH ATTACHED BOXING GLOVE ①, STRIKES AND STARTLES OWL INTO FLIGHT ②, WHICH IS NORMALLY SOUND ASLEEP ③, PULLING STRING ④, WHICH WISKS BROOM ⑤, LEFT AND RIGHT, WHICH JERKS ANOTHER STRING ⑥, WHICH WAVES ROD ⑦, AND SUSPENDED PORK CHOP ⑧, IN FRONT OF POOCH'S NOSE ⑨, WHICH CAUSES POOCH TO DRAG SWEEPER FORWARD.

few, are examples of organizations that need mechanical engineers. Experienced engineers are needed in the missile and space industries in the design and development of such items as gas turbine compressors and power plants, air-cycle cooling turbines, electrically and hydraulically driven fans, and high-pressure refrigerants. Mechanical engineers are also needed in the development and testing of airborne and missile fuel systems, servovalves, and mechanical–electrical control systems. In addition, an engineer may be employed for research endeavor as a university professor, or in the governments of cities, states, and the nation.

Metallurgical engineering and materials science

In many respects the past 25 years may be said to be an "age of materials"—an age which has seen the maturing of space exploration, nuclear power, digital computer technology, and ocean conquest. None of these engineering triumphs could have been achieved without the contributions of the metallurgical engineer. Metals are found in every part of the earth's crust, but rarely in immediately usable form. It is the metallurgical engineer's job to separate them from their ores and from other materials with which they exist in nature.

Metallurgical engineering may be divided into two branches. One branch deals

Illustration 3-11
Almost every aspect of our life is affected by advances in metallurgy and materials science. For example, teeth can now be straightened because of the development of a special type of steel . . . rustproof, strong yet ductile, and hard yet smooth . . . unchanged through ice-cold sodas and red-hot pizzas.

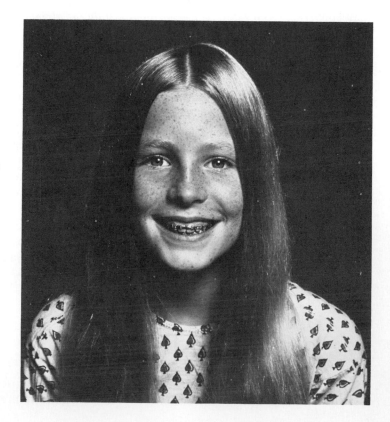

Figure 3-11

"Don't you think you're carrying metallurgical engineering a little too far?"

with the location and evaluation of deposits of ore, the best way of mining and concentrating the ore, and the proper method of refining the ore into the basic metals. The other branch deals with the fabrication of the refined metal or metal alloy into various machines or metal products.

The metallurgist performs pure and applied research on vacuum melting, arc melting, and zone refining to produce metallic materials having unusual properties of strength and endurance. In addition the metallurgist in the aircraft and missile industries is called upon to recommend the best materials to use for special applications, and is frequently called on to give an expert opinion on the results of fatigue tests of metal parts of machines.

The engineer who has specialized in materials science is in great demand today because of the urgent need for human-made composites—the joining of the two or more different materials for the purpose of gaining advantageous or overcoming disadvantageous characteristics of each.

Mining and geological engineering

The mining and geological engineer of today who searches the earth for hidden minerals is necessarily a person of quite different stature than the traditional explorer

Illustration 3-12
Geological engineers investigate new sources of essential mineral bodies.

Figure 3-12

"Yep Zeb, we've struck it big, but she'll sure be tough to get out . . ."

of yesteryear. These engineers must possess a combination of fundamental engineering and scientific education and field experience to enable them to unravel the story of the earth's crust. They must be expert in utilizing very sensitive instruments as they seek to locate new mineral deposits and to anticipate the problems that might arise in getting them out and transporting them to civilization. For this reason it is not unusual to find a mining or geological engineer in a modern office building in New York one week, and the next in Arizona or Afghanistan—or commuting between an expedition campsite and technical laboratories.

The work of mining and geological engineers lies generally in three areas: finding the ore, extracting it, and preparing the resulting minerals for manufacturing industries to use. They design the mine layout, supervise the construction of mine shafts and tunnels in underground operations, and devise methods for transporting minerals to processing plants. Mining engineers are also responsible for mine safety and the efficient operation of the mine, including ventilation, water supply, power, communications, and equipment maintenance. Geological engineers are more directly concerned with locating and appraising mineral deposits.

An important part of the mining and geological engineer's work is to keep in mind inherent air- and water-pollution problems that might develop during the mining operation. This involves establishing efficient controls to prevent harmful side effects of mining and designing ways whereby the land will be restored for people to use after the mining operation terminates.

Naval architecture and marine engineering

For many centuries the sea has played a dominant role in the lives of peoples of all cultures and geographical locations. For this reason in every era the designers of ships have been held in the highest regard for their knowledge and understanding of the sea's physical influences and for their artistry and ability in marine craftmanship. As our civilization increases in complexity, all peoples of future generations will depend to an even greater extent upon vessels of the sea to keep food, materials, and fuel flowing.

Ship design is a refined art as well as an exacting science since most ships are custom built—one at a time. Many large ships are virtually floating cities containing their own power sources, sanitary facilities, food preparation center, and recreational and sleeping accommodations. Every service that would be provided to city dwellers must also be provided for the ship's crew. As with aircraft design, the ship's structural members and intricate networks of piping and electrical circuits must fit together harmoniously in the minimum space possible.

Marine engineers must have a broad-based engineering educational background. They will routinely establish specifications for the vessel, perform design calculations, and install and test ship machinery.

The basic design of seaworthy cargo ships has changed very little from 1900 to 1960. However, with the advent of nuclear power and sophisticated electronic computers a new era in ship design has begun to develop. Surface-effect or air-cushion-type vehicles, submarine tankers, and deep-submergence research vehicles

Illustration 3-13
The marine engineer's role is significant to the economy and defense of the nation.

Figure 3-13

*"But Mr. Plunge, we didn't know you were going to hit it with a **non-breakable** bottle . . . !"*

have all emerged from the realm of science fiction to enter one of engineering reality. The application of these newer ideas for the shipbuilding industry awaits only a more positive commitment to the task by government and industry. With the shrinking world supply of food and energy, this commitment is certain to come.

Nuclear engineering

Nuclear engineering is one of the newest and most challenging branches of engineering. Although much work in the field of nucleonics at present falls within the realm of pure research, a growing demand for people educated to utilize recent discoveries for the benefit of humankind has led many colleges and universities to offer courses in nuclear engineering. The nuclear engineer is familiar with the basic principles involved in both fission and fusion reactions; and by applying fundamental engineering concepts, he is able to direct the enormous energies involved in a proper manner. Work involved in nuclear engineering includes the design and operation of plants to concentrate nuclear reactive materials, the design and operation of plants to utilize heat energy from reactions, and the solution of problems arising in connection with safety to persons from radiation, disposal of radioactive wastes, and decontamination of radioactive areas.

Illustration 3-14
These nuclear engineers are preparing a 10-g sample of neptunium-237 for a replacement measurement in the natural uranium-metal-reflected critical assembly.

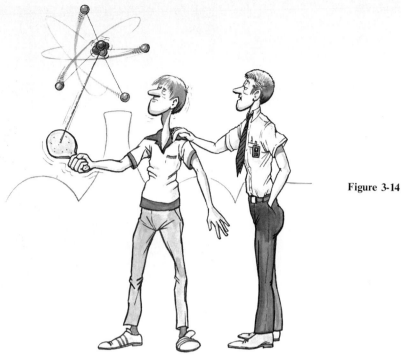

Figure 3-14

"You're right Sharpley . . . I think you've finally got it!"

The wartime uses of nuclear reactions are well known, but of even more importance are the less spectacular peacetime uses of controlled reactions. These uses include such diverse applications as electrical power generation and medical applications. Other applications are in the use of isotopes in chemical, physical, and biological research, and in the changing of the physical and chemical properties of materials in unusual ways by subjecting them to radiation.

Recent advances in our knowledge of controlled nuclear reactions have enabled engineers to build power plants that use heat from reactions to drive machines. Submarine nuclear power plants, long a dream, are now a reality, and experiments are being conducted on smaller nuclear power plants that can be used for airborne or railway applications.

At present, ample opportunities for employment of nuclear engineers exist in both privately owned and government-operated plants, where separation, concentration, or processing of nuclear materials is performed. Nuclear engineers are also needed by companies that may use radioactive materials in research or processing involving agricultural, medical, metallurgical, and petroleum products.

Petroleum engineering

Throughout history the energy available to us beyond our own muscle power has been a measure of hope for a more secure and improved material life. In early Greek and Roman civilizations wind and water provided much of human energy

Illustration 3-15
One of the tasks of the petroleum engineer is to locate oil deposits and to devise methods for oil recovery. As is the case in this photograph, design calculations frequently must be made at the drilling site.

needs. In early America, wood was the primary source of energy. Today the major source of energy is petroleum. It is the most widely used of all energy sources because of its mobility and flexibility in utilization. Approximately three fourths of the total energy needs of the United States are currently supplied by petroleum products, and this condition will likely continue for many years. Petroleum engineering is the practical application of the basic sciences (primarily chemistry, geology, and physics) and the engineering sciences to the development, recovery, and field processing of petroleum.

Petroleum engineering deals with all phases of the petroleum industry, from the location of petroleum in the ground to the ultimate delivery to the user. Petroleum products play an important part in many phases of our everyday life in providing our clothes, food, work, and entertainment. Because of the complex chemical structure of petroleum, we are able to make an almost endless number of different articles. Owing to the wide demand for petroleum products, the petroleum engineer strives to satisfy an ever-increasing demand for oil and gas from the ground.

The petroleum engineer is concerned first with finding deposits of oil and gas in quantities suitable for commercial use, in the extraction of these materials from the ground, and the storage and processing of the petroleum above ground. The petroleum engineer is also concerned with the location of wells in accordance with the findings of geologists, the drilling of wells and the myriad problems associated with the drilling, and the installation of valves and piping when the wells are completed. In addition to the initial tapping of a field of oil, the petroleum engineer is concerned with practices that will provide the greatest recovery of the oil, considering all possible factors that may exist many thousand feet below the surface of the earth.

After the oil or gas has reached the surface, the petroleum engineer will provide the means of transporting it to suitable processing plants or to places where it will be

Figure 3-15

"... Hey kid, that's **not** the way we cap a well!"

used. Pipelines are providing an ever-increasing means of transporting both oil and gas from field to consumer.

Many challenges face the petroleum engineer. Some require pioneering efforts, such as with the rapidly developing Alaska field. Other opportunities lie closer at hand. For example, it is known that because of excessive costs in recovery less than one half of the oil already discovered in the United States *has yet to be brought to the surface of the earth.* It is estimated that even a 10 per cent increase in oil recovery would produce 3 billion barrels of additional oil, a worth of over 20 billion dollars.

Owing to the expanding uses for petroleum and its products, the opportunities for employment of petroleum engineers are widespread. Companies concerned with the drilling, producing, and transporting of oil and gas will provide employment for the majority of engineers. Because of the widespread search for oil, employment opportunities for the petroleum engineer exist all over the world; and for the young person wishing a job in a foreign land, oil companies have crews in almost every country over the globe. Other opportunities for employment exist in the field of technical sales, research, and as civil service employees of the national government.

The curriculum in petroleum engineering includes courses in drilling methods, engines, oil and gas recovery, storage and transportation, and geology.

Problems

3-1. Discuss the changing requirements for aerospace and astronautical engineers.
3-2. Investigate the opportunities for employment in agricultural engineering. Discuss your findings.
3-3. Write a short essay on the differences in the utilization and capability of the architectural engineer and the civil engineer who has specialized in structural analysis.

3-4. Interview a chemical engineer. Discuss the differences in his work and that of a chemist.

3-5. Assume that you are employed as an electrical engineer. Describe your work and comment particularly concerning the things that you most like and dislike about your job.

3-6. Explain why the demand for industrial engineers has increased significantly during the past ten years.

3-7. Write a 200-word essay describing the challenging job opportunities in engineering that might be particularly attractive for an engineering graduate.

3-8. Explain the importance of mechanical engineers in the electronics industry.

3-9. Describe the changes that might be brought about to benefit mankind by the development of new engineering materials.

3-10. Investigate the need for nuclear and petroleum engineers in your state and report on your findings.

4

Work opportunities in engineering

During the college years engineering students will study courses in many subject areas. Language courses make it easier to organize and present ideas effectively; mathematics courses stress the manipulation of symbols as an aid in problem solving; social science courses assist in finding a place in society as informed citizens; and various technical courses help to gain an understanding of natural laws. In your study of technical courses, you will become familiar with a store of factual information that will form the basis for your engineering decisions. The nature of these technical courses, in general, will influence your choice of a major field of interest. For example, you may decide to concentrate your major interest in some particular field such as civil, chemical, electrical, or mechanical engineering.

The college courses also provide training in learning facts and in developing powers of reasoning. Since it is impossible to predict what kind of work a practicing engineer will be doing after graduation, the objective of an engineering education is to provide a broad base of facts and skills upon which the engineer can rely.

It usually is not sufficient to say that an engineer is working as a *civil engineer*. The work experience may vary over a wide spectrum. As a civil engineer, for example, one may be performing research on materials for surfacing highways, or be employed in government service and be responsible for the budget preparation of a missile launch project. In fact, there are many things that a practicing engineer will

Research is an organized method of trying to find out what you are going to do after you find that you cannot do what you are doing now.
—Charles F. Kettering

be called upon to do which are not described by his or her major course of study. The *type* of work that the engineer may do, as differentiated from a major field of specialization, can be called "engineering function." Some of these functions are research, development, design, production, construction, operations, sales, and management.

It has been found that in some engineering functions, such as in the management of a manufacturing plant, specialization is of lesser importance, whereas in other functions, such as research in transistor theory, specialization may be extremely important. In order to understand more fully the activities of a practicing engineer, let us examine some of these functions.

Research

In some respects research is one of the more glamorous functions of engineering. In this type of work the engineer delves into the nature of matter, exploring processes to use engineering materials and searching for reasons for the behavior of the things that make up our world. In many instances the work of the scientist and the engineer who are engaged in research will overlap. The work of scientists usually is closely allied with research. The objective of the research scientist is to *discover truths*. The

Illustration 4-1
Research is an important type of work performed by the engineer. In research he employs basic scientific principles in the discovery and application of new knowledge. This engineer is experimenting with a wind tunnel model of the city of Denver, Colorado, during a simulated atmospheric inversion.

objective of the research engineer, on the other hand, usually is directed toward the practical side of the problem: not only to discover but also *to find a use for the discovery*.

The research engineer must be especially perceptive and clever. Patience is also required, since most tasks have never before been accomplished. It is also important to be able to recognize and identify phenomena previously unnoticed. As an aid to training an engineer to do research work, some colleges give courses in research techniques. However, the life of a research engineer can be quite disheartening. In addition to probing and exploring new areas, much of the work is trial and error, and outstanding results of investigation usually occur only after long hours of painstaking and often discouraging work.

Until within the last few decades, almost all research was solo work by individuals. However, with the rapid expansion of the fields of knowledge of chemistry, physics, and biology, it became apparent that groups or "research teams" of scientists and engineers could accomplish better the aims of research by pooling their efforts and knowledge. Within the teams, the enthusiasm and competition provide added incentive to push the work forward, and since each person is able to contribute from a particular specialty, discovery is accelerated.

As has been indicated, a thorough training in the basic sciences and mathematics is essential for a research engineer. In addition, an inquiring mind and a great curiosity about the behavior of things is desirable. Most successful research engineers have a fertile and uninhibited imagination and a knack of observing and questioning phenomena that the majority of people overlook. For example, one successful research engineer has worked on such diverse projects as an automatic lawnmower, an electronic biological eye to replace natural eyes, and the use of small animals as electrical power sources.

Most research engineers secure advanced degrees because they need additional training in basic sciences and mathematics, and, in addition, this study usually gives them an opportunity to acquire useful skills in research procedures.

Development

After a basic discovery in natural phenomena is made, the next step in its utilization involves the development of processes or machines that employ the principles involved in the discovery. In the research and development fields, as in many other functions, the areas of activity overlap. In many organizations the functions of research and development are so interrelated that the department performing this work is designated simply as a research and development (R and D) department.

The engineering features of development are concerned principally with the actual construction, fabrication, assembly, layout, and testing of scale models, pilot models, and experimental models for pilot processes or procedures. Where the research engineer is concerned more with making a discovery that will have commercial or economic value, the development engineer will be interested primarily in producing a process, an assembly, or a system *that will work*.

The development engineer does not deal exclusively with new discoveries. Actually, the major part of work assignments will involve using well-known principles and employing existing processes or machines to perform a new or unusual

Illustration 4-2
Time spent in library research can be very rewarding in the saving of both time and money by preventing unnecessary design and development work.

function. It is in this region that many patents are granted. In times past, the utilization of basic machines, such as a wheel and axle, and fundamental principles, including Ohm's Law and Lenz' Law, have eventually led to patentable articles, such as the electric dynamo. On the other hand, within a very short time after the announcement of the discovery of the laser in 1960, a number of patents were issued on devices employing this new principle. Thus the lag between the discovery of new knowledge and the use of that knowledge has been steadily decreasing through the years.

In most instances the tasks of the development engineer are dictated by immediate requirements. For example, a new type of device may be needed to determine at all times the position in space of an airplane. Let us suppose that the development engineer does not know of any existing device that can perform the task to the desired specifications. Should he or she immediately attempt to invent such a device? The answer, of course, is "usually not." First, the files of available literature should be searched for information pertaining to existing designs (Illustration 4-2).

Such information may come from two principal sources. The first source is library material on processes, principles, and methods of accomplishing the task or related tasks. The second source is manufacturers' literature. It has been said humorously that there is no need to reinvent the wheel. A literature search may reveal a device that can accomplish the task with little or no modification. If no device is available that will do the work, a system of existing subassemblies may be set up and joined to accomplish the desired result. Lacking these items, the development engineer must go further into basic literature, and, using results from experiments throughout the world, formulate plans to construct a model for testing. Previous research points a way to go, or perhaps a mathematical analysis will provide clues as to possible methods.

The development engineer usually works out ideas on a trial or "breadboard" basis, whether it be a machine or a computer process. Having the parts or systems somewhat separated facilitates changes, modifications, and testing. In this process, improved methods may become apparent and can be incorporated. When the system or machine is in a workable state, the development engineer must then refine it and package it for use by others. Here again ingenuity and a knowledge of human nature are important. A device that works satisfactorily in a laboratory when manipulated by skilled technicians may be hopelessly complex and unsuited for field use. The development engineer is the important person behind every pushbutton.

Illustration 4-3
Development engineers use the results of basic research and convert them into models and prototypes for full-scale testing and evaluation. In this picture, a team of engineers and technicians are shown preparing a test to evaluate the results of a head-on automobile crash into an immovable barrier.

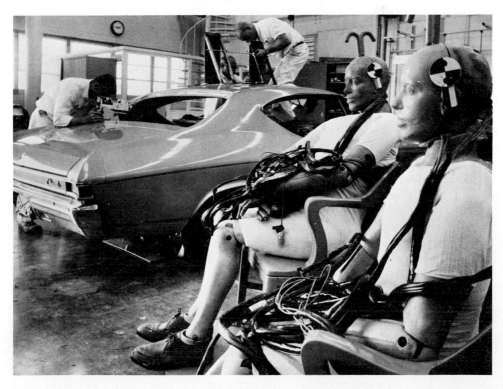

The training of an engineer for development work is similar to the training that the research engineer will expect to receive. However, creativity and innovation are perhaps of more importance, since the development engineer is standing between the scientist or the research engineer and the members of management who provide money for the research effort. The economic value of certain processes over others to achieve a desired result must be recognizable. It is also important to be able to convince others of the soundness of any conclusions reached. A comprehensive knowledge of basic principles of science and an inherent cleverness in making things work are also essential skills for the development engineer.

Design

In our modern way of life, mass production has given us cheaper products and has made more articles available than ever before in history. In the process of producing these articles, the design engineer enters the scene just before the actual manufacturing process begins. After the development engineer has assembled and tested a device or a process and it has proved to be one that it is desirable to produce for a mass market, the final details of making it adaptable for production will be handled by a design engineer.

In bridging the gap between the laboratory and the production line, the design engineer must be a versatile individual. This requires a mastery of basic engineering principles and mathematics, and an understanding of the capabilities of machines. It is also important to understand the temperament of the people who operate them. The design engineer must also be conscious of the relative costs of producing items, for it will be the design that will determine how long the product will survive in the open market. Not only must the device or process work, it must also be made in a style and at a price that will attract customers.

As an example, let us take a clock, a simple device widely used to indicate time. It includes a power source, a drive train, hands, and a face. Using these basic parts, engineers have designed spring-driven clocks, weight-driven clocks, and electrically driven clocks with all variations of drive trains. The basic hands and face have been modified in some models to give a digital display (Illustration 4-4). The case has been made in many shapes and, perhaps in keeping with the slogan "time flies," it has even been streamlined! In the design of each modification the design engineer has determined the physical structure of the assembly, its aesthetic features, and the economics of producing it.

Of course the work of the design engineer is not limited solely to performing engineering on mass-produced items. Design engineers may work on items such as bridges or buildings in which only one of a kind is to be made. However, in such work they are still fulfilling the design process of adapting basic ideas to provide for making a completed product for the use of others. In this type of design engineers must be able to use their training, in some cases almost intuitively, to arrive at a design solution which will provide for adequate safety without excessive redundancy. The more we learn about the behavior of structural materials, the better we can design without having to add additional materials to cover the "ignorance factor" area. Particularly in the aircraft industry, design engineers have attempted to use structural materials with minimum excess being allowable as a safety factor. Each part must perform without failure, and every ounce of weight must be saved.

Illustration 4-4
*The availability of solid-state
circuitry has revolutionized
the design of clocks and
watches.*

Of course to do this, fabricated parts of the design must be tested and retested for resistance to failure due either to static loads or to vibratory fatiguing loads. Also, since surface roughness has an important bearing on the fatigue life of parts which are subjected to high stress or repeated loads, much attention must be given to specifying in designs that surface finishes must meet certain requirements.

Since design work involves a production phase, the design engineer is always considering costs as a factor in our competitive economy (Illustration 4-5a and b). One way in which costs can be minimized in manufacture or construction is to use standard parts, and standard sizes and dimensions for raw material. For example, if a machine were designed using nonstandard bolt threads or a bridge designed using nonstandard steel I-beams, the design probably would be more expensive than needed to fulfill its function. Thus, the design engineer must be able to coordinate the parts of a design so that it functions acceptably and is produced at minimum cost.

The design engineer soon comes to realize also that there usually is more than one acceptable way to solve a design problem. Unike an arithmetic problem with fixed numbers which give one answer, real design problems can have many answers and many ways of obtaining a solution, *and all may be acceptable.* In such a case the engineer's decision becomes a matter of experience and judgment. At other times it may become just a matter of making a decision one way or the other. Regardless of the method used, the final solution to a problem should be a conscious effort to provide the *best* method, considering fabrication, costs, and sales.

What are the qualifications of a design engineer? Creativity is a key element. Every design will embody a departure from what has been done before. However, all designs must be produced within the constraints of the reality of the physical properties of materials and by economic factors. Therefore, design engineers must be thoroughly knowledgeable in fundamental engineering in a rather wide range of subjects. In addition, they must be familiar with basic principles of economics, both from the standpoint of employing people and using machines. As they progress upward into supervisory and management duties, the employment of principles of psychology and economics becomes of even more importance. For this reason they usually will have more use for management courses than will research or development engineers.

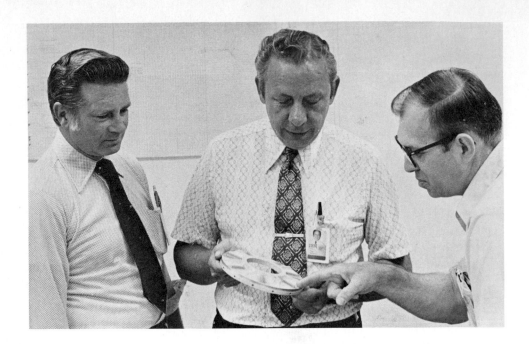

Illustration 4-5a and 4-5b
The design engineers pictured in 4-5a are evaluating a cost-reduction proposal for the connector base plate assemby shown in 4-5b.

Production and construction

In the fields of production and construction, the engineer is more directly associated with the technician, mechanic, and laborer. The production or construction engineer must take the design engineer's drawings and supervise the assembly of the object as it was conceived and illustrated by drawings or models.

Illustration 4-6
Construction engineers are responsible for seeing that a project is carried out as designed. Usually, they work with large projects, where weather and terrain are complicating factors.

Usually, production or construction engineers are associated closely with the process of estimating and bidding for competitive jobs. In this work they employ their knowledge of structural materials, fabricating processes, and general physical principles to estimate both time and cost to accomplish tasks. In construction work the method of competitive bidding is usually used to award contracts, and the ability to reduce an appropriate amount from an estimate by skilled engineering practices may mean the difference between a successful bid and one that is either too high or too low.

Once a bid has been awarded, it is usual practice to assign a "project engineer" as the person who assumes overall responsibility and supervision of the work from the standpoint of materials, labor, and money. The engineer will supervise other production or construction engineers, who will be concerned with more specialized features of the work, such as civil, mechanical, electrical, or chemical engineering. Here the project engineer must complete the details of the designers' plans. Provision must be made to provide the specialized construction tools needed for the work. Schedules of production and/or construction must also be set up and questions that technicians or workers may raise concerning features of the design must be answered. Design engineers will need to be advised concerning desirable modifications that will aid in the construction or fabrication processes. In addition, the project engineer must be able to work effectively with construction or production crafts and labor unions.

Preparation of a schedule for production or construction is an important task of the engineer. In the case of an industrial plant, all planning for the procurement of raw materials and parts will be based upon this production schedule. An assembly line in a modern automobile manufacturing plant is one example which illustrates the necessity for scheduling the arrival of parts and subassemblies at a predetermined time. As another example, consider the construction of a modern multistory office building. The necessity for parts and materials to arrive at the right time is very important. If they arrive too soon, they probably will be in the way, and if they arrive too late, the building is delayed, which will cause an increase in costs to the builder.

Qualifications for production or construction engineers includes a thorough knowledge of basic engineering principles. In addition, they must have the ability to visualize the parts of an operation, whether it be the fabrication of a solid-state computer circuit or the building of a concrete bridge. From an understanding of the operations involved, they must be able to arrive at a realistic schedule of time, materials, and manpower. Therefore, emphasis should be placed upon courses in engineering design, economics, business law, and psychology.

Operations

In modern industrial plants, the number and complexity of machines, the equipment and buildings to be cared for, and the planning needed for expansion have brought out the need for specialized engineers to perform services in these areas. If a new manufacturing facility is to be constructed, or an addition made to an existing facility, it will be the duty of a plant engineer to perform the basic design, prepare the proposed layout of space and location of equipment, and to specify the fixed equipment such as illumination, communication, and air conditioning. In some cases, the work of construction will be contracted to outside firms, but it will be the general responsibility of the plant engineer to see that the construction is carried on as he has planned it.

After a building or facility has been built, the plant engineer and an appropriate staff are responsible for maintenance of the building, equipment, grounds, and utilities. This work varies from performing routine tasks to setting up and regulating the most complex and automated machinery in the plant.

The plant engineer must have a wide knowledge of several branches of engineering in order to perform these functions. For land acquisition and building construction, civil engineering courses will be needed. For working with equipment and machinery, mechanical engineering courses are needed. For work with power generation equipment, mechanical and electrical backgrounds are essential. For work in specialized parts of the plant, knowledge is needed in such fields as chemical, metallurgical, nuclear, petroleum, or textile engineering.

In many plants, particularly in utility plants, the engineer also is concerned with operation of the plant. Boilers, generators, turbines, and accessory equipment must be operated at their best efficiency. Plant engineers should be able to compare costs of operating under various conditions, and set schedules for machines so that the best use will be made of them. In the case of chemical plants, they will also attempt to regulate the flows and temperatures at levels that will produce the greatest amount of desired product at the end of the line.

In the dual role as a plant and operations engineer, it is important to evaluate new equipment as it becomes available to see whether additional operating econo-

Illustration 4-7
The plant engineer is responsible for the operation and maintenance of the equipment, buildings, grounds, and utilities.

mies can be secured by retiring old equipment and installing new types. In this function the engineer must frequently assume a salesperson's role in order to convince management that it should discard equipment that, apparently, is operating perfectly and spend money for newer models. Here the ability to combine facts of engineering and economics is invaluable.

Plant engineering, of course, will be associated closely with production engineering processes. The production engineer will create needs for new machines, new facilities, and new locations. The plant engineer will correlate such things as the building layout, machine location, power supplies, and materials handling equipment so that they best will serve the needs of production.

The general qualifications of plant and operations engineers have already been mentioned. They must have basic knowledge of a wide variety of engineering fields such as civil, chemical, electrical, and mechanical, and also they must have specialized knowledge of areas peculiar to their plant and its operation. In addition, plant and operations engineers must be able to work with people and machines and to know what results to expect from them. In this part of their work, a knowledge of industrial engineering principles is valuable. In addition, it is desirable to have a basic understanding and knowledge of economics and business law. In this work, in general, training in detailed research procedures and abstract concepts is of lesser importance.

Sales

An important and sometimes unrecognized function in engineering is the realm of applications and sales. As is well known, the best designed and fabricated product is of little use unless a demand either exists or has been created for it. Since many new processes and products have been developed within the past few years, a field of work has opened up for engineers in presenting the use of new products to prospective customers.

Discoveries and their consequent application have occurred so rapidly that a product may be available about which even a recent graduate may not know. In this case, it will fall to the engineer in sales who has intimate knowledge of the principles involved to go out and educate possible users so that a demand can be created. In this work the engineer must assume the role of a teacher. In many instances the product must be presented primarily from an engineering standpoint. If the audience is composed of engineers, "talk their language" and answer their technical questions. But, if the audience includes nonengineers, present the features of the product in terms that can be understood.

In addition to acquiring a knowledge of the engineering features of a particular product, the sales and application engineer must also be familiar with the operations of the customer's plant. This is important from two standpoints. First, you should be able to show how your product will fit into the plant, and also you must show the economics involved to convince customers that they should buy it. At the same time,

Illustration 4-8
In sales, the engineer must be able to describe a technical product to customers and show how they will benefit from using the latest developments.

you must point out the limitations of your product and the possible changes necessary to incorporate it into a new situation. For example, a new bonding material may be available, but in order for a customer to use it in an assembly of parts, a special refrigerator for storage may be necessary. Also, the customer would need to have emphasized the necessity for proper cleaning and surface preparation of the parts to be bonded.

A second reason that the sales and application engineer must be familiar with a customer's plant operation is that many times new requirements are generated here. By finding an application area in which no apparatus is available to do the work, the sales engineer is able to report back to the company that a need exists and that a development operation should be undertaken to produce a device or process to meet the need.

Almost all equipment of any complexity will need to be accompanied by introductory instructions when it is placed in a customer's plant. Here the application engineer can create goodwill by conducting an instruction program outlining the capabilities and limitations of the equipment. Also, after the equipment is in service, maintenance and repair capabilities by competent technical personnel will serve to maintain the confidence of customers.

Sales and applications engineers should have a basic knowledge of engineering principles and should, of course, have detailed knowledge in the area of their own products. Here the ability to perform detailed work on abstract principles is of less importance than the ability to present one's ideas clearly. A genuine appreciation of people and a friendly personality are desirable personal attributes. In addition to basic technical subjects, courses in psychology, sociology, and human relations will prove valuable to the sales and applications engineer.

Usually, an engineer will spend several years in a plant learning the processes and the details of the plant's operation and management policies before starting out to be a member of the sales staff. As a sales engineer you will represent your company in the mind of the customer. Therefore, you must present a pleasing appearance and give a feeling of confidence in your engineering ability.

Management

Results of recent surveys show that the trend today is for corporate leaders in the United States to have backgrounds in engineering and science. In a survey of some 600 large industrial firms, 20 technical and engineering colleges and universities have four or more of their graduates serving as board chairmen, presidents, or senior vice-presidents in these firms.

It has been predicted that within ten years, the *majority* of corporation executives will be persons who are trained in engineering and science as well as in business and the humanities, and who can bridge the gap between these disciplines.

Since the trend is toward more engineering graduates moving into management positions, let us examine the functions of an engineer in management.

The basic functions of the management of a company are generally similar whether the company objective is dredging for oyster shells or building diesel locomotives or digital computers. These basic functions involve using the capabilities of the company to the best advantage to produce a desirable product in a competitive economy. The use of the capabilities, of course, will vary widely depending upon the enterprise involved.

The executive of a company, large or small, has the equipment in the plant, the labor force, and the financial assets of the organization to use in conducting the plant's operations. In management, the engineer must make decisions involving all three of these items.

In former years it was assumed that only persons trained and educated in business administration should aspire to management positions. However, now it has been recognized that the education and other abilities which make a good engineer also provide the background to make a good management executive. The training for correlating facts and evaluating courses of action in making engineering decisions can be carried over to management decisions on machinery, personnel, and money. In some cases, the engineer is technically strong but may be quite naïve in the realm of business practicability. Therefore, it is in the business side of an operation that the engineer usually must work harder to develop the necessary skills.

The engineer in management is concerned more intimately with the long-range effects of policy decisions. Where the design engineer considers first the technical phases of a project, the engineer in management must consider how a particular decision will affect the employees who work to produce a product and how the decision will affect the people who provide the financing of the operation. It is for

Illustration 4-9
The engineer in management must have the ability to reduce a large number of variables to the most significant factors and then move decisively to a plan of action.

this reason that the management engineer is concerned less with the technical aspects of the profession and relatively more with the financial, legal, and labor aspects.

This does not imply that engineering aspects should be minimized or deleted. Rather the growing need for engineers in management shows that the type and complexity of the machines and processes used in today's plants require a blending of technical and business training in order to carry forward effectively. Particularly is this trend noted in certain industries, such as aerospace and electronics, where the vast majority of executive managerial positions are occupied by engineers and scientists. As other industries become automated, a similar trend in those fields also will become apparent.

The education that an engineer in management receives should be identical to the basic engineering education received in other engineering functions. However, a young engineers usually can recognize early in their careers whether or not they have an aptitude for working with people and directing their activities. If you have the ability to "sell your ideas" and to get others to work with you, probably you can channel your activities into managerial functions. You may start out as a research engineer, a design engineer, or a sales engineer, but the ability to influence others to your way of thinking, a genuine liking for people, and a consideration for their responses, will indicate that you probably have capabilities as a manager.

Of course, management positions are not always executive positions, but the ability to apply engineering principles in supervisory work involving large numbers of men and large amounts of money is a prerequisite in management engineering.

Other engineering functions

A number of other engineering functions can be considered that do not fall into the categories previously described. Some of these functions are testing, teaching, and consulting.

As in the other functions, there are no specific curricula leading directly toward these types of work. Rather a broad background of engineering fundamentals is the best guide to follow in preparing for work in these fields.

In testing, the work resembles design and development functions most closely. Most plants maintain a laboratory section that is responsible for conducting engineering tests of proposed products or for quality control on existing products. The test engineer must be qualified to follow the intricacies of a design and to build suitable test machinery to give an accelerated test of the product. For example, in the automotive industry, not only are the completed cars tested, but also individual components, such as engines, brakes, and tires, are tested on stands to provide data to improve their performance. The test engineer must be able also to set up quality control procedures for production lines to ensure that production meets certain standards. In this work, mathematics training in statistical theory is helpful.

A career in teaching is rewarding for many persons. A desire to help others in their learning processes, a concern for some of their personal problems, and a thorough grounding in engineering and mathematics are desirable for those considering teaching engineering subjects. In the teaching profession, the trend today is toward the more theoretical aspects of engineering, and a person will usually find that teaching is more closely allied with research and development functions than with others. Almost all colleges now require the faculty to obtain advanced degrees, and a person desiring to be an engineering teacher should consider seriously the desirability of obtaining a doctorate in his or her chosen field.

More and more engineers are going into consulting work. Work as an engineering consultant can be either part time or full time. Usually, a consulting engineer is a person who possesses specific skills in addition to several years of experience, and may offer services to advise and work on engineering projects either part time or full time.

Frequently, two or more engineers will form an engineering consulting firm that employs other engineers, technicians, and draftsmen and will contract for full engineering services on a project. The firm may restrict engineering work to rather narrow categories, such as the design of irrigation projects, power plants, or aerospace facilities, or a staff may be available that is capable of working on a complete spectrum of engineering problems.

Illustration 4-10
Teaching is a rewarding activity of engineering. Frequently the engineering professor is the first person to introduce the student to the ethics and responsibilities of the profession.

On the other hand, as a consulting engineer you may prefer to operate alone. Your firm may consist of only your own skills such that, in a minimum time, you may be able to advise and direct an operation to overcome a given problem. For instance, you may be employed by an industrial plant. In this way the plant may be able to solve a given problem more economically, particularly if the required specialization is seldom needed by the plant.

As may be inferred, a consulting engineer must have *specific* skills to offer, and must be able to use his or her creative ability to apply individual skills to unfamiliar situations. Usually, these skills and abilities are acquired only after several years of practice and postgraduate study.

Consulting work is an inviting part of the engineering profession for a person who desires self-employment, and is willing to accept its business risks to gain an opportunity for financial reward.

Engineering functions in general

As described in previous paragraphs, training and skills in all functions are basically the same, that is, fundamental scientific knowledge of physical principles and mathematics. However, it can be seen that research on one hand and management on the other require different educational preparations.

For work in research, emphasis is on theoretical principles and creativity, with little emphasis on economic and personnel considerations. On the other hand, in management, primary attention is given to financial and labor problems and relatively little to abstract scientific principles. Between these two extremes, we find the other functions with varying degrees of emphasis on research- or managerial-oriented concepts.

Figure 4-1 shows an idealized image of this distribution. Bear in mind that this diagram merely depicts a trend and does not necessarily apply to specific instances.

To summarize the functions of the engineer, we can say that in all cases he or she is a problem identifier and solver. Whether it be a mathematical abstraction that may have an application to a nuclear process or a meeting with a bargaining group at a conference table, it is a problem that must lie identified and reduced to its essentials and the alternatives explored to reach a solution. The engineer then must apply specialized knowledge and inventiveness to select a reasonable method to achieve a result, even in the face of vague and sometimes contradictory data. That the engineer has been able, in general, to accomplish this is proven by a long record of successful industrial management and productivity.

Problems

4-1. Discuss an important scientific breakthrough of the past year that was brought about by an engineering research effort.
4-2. Discuss the differences between engineering research and engineering development.

Figure 4-1 Application of principles in various engineering functions.

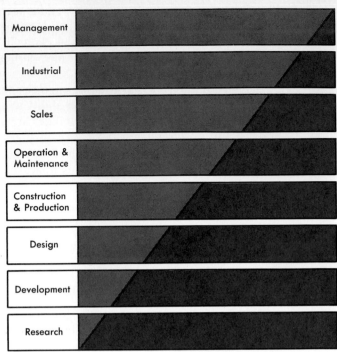

FINANCIAL PRINCIPLES & MANPOWER UTILIZATION

Management

Industrial

Sales

Operation & Maintenance

Construction & Production

Design

Development

Research

ABSTRACT SCIENTIFIC PRINCIPLES

4-3. Interview an engineer and estimate the percentage of his or her work that is devoted to research, development, and design.

4-4. Discuss the importance of the engineer's design capability in modern industry.

4-5. Investigate the work functions of the engineer and write a brief essay describing the function that most appeals to you.

4-6. Discuss the importance of the sales engineer in the total engineering effort.

4-7. Interview an engineer in management. Discuss the reasons that many engineers rise to positions of leadership as managers.

4-8. Compare the engineering opportunities in teaching with those in industry.

4-9. Investigate the opportunities for employment in a consulting engineering firm. Discuss your findings.

4-10. Discuss the special capabilities required of the engineer in construction.

5

Professional responsibilities of the engineer

The word "professional" is used in many ways and has many meanings. It can be used in the sense of the skill of a professional athlete who receives pay for his or her efforts, as distinguished from an amateur who performs more for the joy of performing (Figure 5-1). It can be used in the sense of a type of work, as in describing a professional job of house painting done by an experienced painter. Also, it can be used merely to describe a degree of effort or line of conduct over a period of time, as used in the expression "a professional beggar." However, in the sense that engineers would employ the word "professional," it should be restricted to a particular and specialized group of people, identified by distinguishing characteristics that separate its members from nonprofessionals.

Within the last century, three groups have emerged with the title "learned professions." These professional groups are law, medicine, and theology. These groups came into being gradually over a long period of time and had certain characteristics in common, among which were higher levels of educational achievement and a sincere desire for performing a service for people. There is no formal naming of a person or group of persons to professional status, nor is there a schedule or procedure to follow to achieve recognition as a professional. Rather the group itself sets standards of training, skills, achievement, and service in order to call itself a professional group, and the public accepts the group's evaluation of itself.

Who is a professional? As generally used in the sense of the learned professions, a professional person is one who applies certain knowledge and skill, usually obtained by college education, for the service of people. In addition, a professional person observes an acceptable code of conduct, uses discretion and judgment in dealing with people, and respects their confidences. Also, professional persons usually have legal status, use professional titles, and associate together in groups. Although engineering has met most of these criteria for a long time, it has been only within the last few decades that legal status has been conferred upon the engineering profession.

Figure 5-1

*"... Perhaps you could tell the audience just
how you decided to make basketball your profession ..."*

The engineer as a
professional person

Knowledge and skill above that of the average person is a characteristic of the
professional. Where a worker will have specific skills in operating a particular machine, a professional person is considered able to apply fundamental principles that
are usually beyond the range of the average worker. The knowledge of these principles as well as the skills necessary to apply them distinguishes the professional. The
engineer, because of an education in the basic sciences, mathematics, and engineering sciences, is capable of applying basic principles for such diverse things as improving the construction features of buildings, developing processes that will provide new
chemical compounds, or designing tunnels to bring water to aid areas.

An important concept in the minds of most persons is that a professional person
will perform a service for people. This means that service must be considered ahead
of any monetary reward that a professional person may receive. In this respect the
professional, individually, should recognize that a need for personal services exists
and seek ways to provide a solution to these needs. Almost all engineering is performed to fill a need in some phase of our society. It may be to develop better

Illustration 5-1
As a professional person the engineer works in a confidential relationship with clients so that proprietary information is protected and trade secrets are not divulged.

appliances for the household, or to provide better transportation facilities, or to make possible a better life in regions of unfavorable climate.

Discretion and judgment also characterize a professional person. In most situations a choice of several methods to accomplish a given task will be available. The engineer must consider the facts available and the principles that apply and make decisions based upon these rather than upon expediency. Consideration must be given not only to the mechanical aspects of a solution but also to the effects that a particular decision will have upon the persons concerned.

A professional person is one in whom confidence can be placed. This confidence is not only in his or her skill and ability but also in that knowledge of a client's business or trade information or personal matters will not be divulged improperly. The engineer works in a relation of confidence with a client or employer not to divulge trade secrets or to take any advantage of any knowledge that may harm the client or employer (Illustration 5-2). The public, in general, will have confidence that the engineer's design of buildings, bridges, or power systems will be adequate and safe to use. The engineer must not fail the public in this responsibility.

All professionals adhere to a code of ethical conduct. This code of ethics outlines the standards to which members of the group subscribe and gives an understanding of what the public can expect in its relationship with the profession. The code of ethics also serves as a guide to the members of the profession in their conduct and relations with each other.

Illustration 5-2
*Engineers must be able
to work at various levels
of confidentiality and
secrecy.*

Legal status usually is a characteristic of a professional. A medical doctor, for example, has certain rights and privileges afforded by law. Legal recognition of a professional group is afforded by a procedure of certification, licensing, or registration. In all states, a registration law is in effect which provides for legal registration of an engineer following submission of evidence of education and technical ability. Registration confers the legal title of "engineer" to the recipient, who may use the initials "P.E." after his or her name to denote registration as a "Professional Engineer."

Professionalism for the engineer

Professionalism is an individual state of mind. It is a way of thinking and living rather than the development of specific skills or the acquiring of certain knowledge. While the mere acquisition of knowledge may make a person more skilled as a clerk or laborer, knowledge alone does not often promote the desire within oneself to serve or be responsive to the needs of people. It is in this realm of service that the engineer joins with members of other learned professional groups in placing honesty and integrity of action above the legal or minimum level allowable (Figure 5-2).

Although knowledge and skill often exist apart from professionalism, professionalism can mature only where such competence creates a proper atmosphere. Where competence is an impersonal quality, professionalism, in contrast, is personal. In addition to a state of mind, it is a way of working and living—a way of adding something valuable to competence. For engineers professionalism implies that maximum use will be made of both skill and knowledge, and that each person will use his or her competence to its fullest extent:

□ With complete honesty and integrity.
□ With his best effort in spite of the fact that frequently neither client nor employer is able to evaluate that effort.
□ With avoidance of all possible conflicts of interest.
□ With the consciousness that the profession of engineering is often judged by the performance of the individual.

Professionalism for engineers must begin with good moral character, because they often occupy positions of trust where they personally must set the standards. Consequently, they may be required to make decisions that differ from the preferences of either company or client.

Figure 5-2

"So it leans a little ... think how much we saved on that spongy land ..."

Professionalism for an engineer means:

☐ Striving to improve all work until it becomes a model for those in the field, as a minimum using the most up-to-date techniques and procedures.
☐ Proper credit for work done and ideas developed by subordinates.
☐ Loyalty to one's employer or client, always with concern for the public safety in construction, product design, plant operation, and all other phases of engineering.
☐ Leadership of less experienced colleagues and subordinates toward personal development and an enthusiasm for the profession.
☐ Activity in technical societies in order to keep current in the field, and encouragement to subordinates to improve their technical competence the same way.
☐ Participation in professional societies, as well as technical societies, thereby demonstrating an interest in the profession and encouraging coworkers to recognize the technical and the professional as of equal-ranking importance.
☐ Registration, not simply because it may be a legal requirement, but more particularly as a demonstration to coworkers and the public that this is one important hallmark of a professional, a willingness to go beyond the minimum to help and encourage others to realize their full potential.

For engineers in various areas of work, professionalism will include special facets that are more particularly related to a particular field. For example, engineers in industry should be especially conscious of their responsibility in protecting "company proprietary" designs or processes. It also means the establishment of performance standards and safety criteria which protect the purchaser while maintaining a satisfactory return to the manufacturer (Illustration 5-3). For the engineer in government or the engineer in private practice, professionalism may mean capitalizing on a special opportunity to project the profession to the public as a constructive force in society. For the engineer in education, professionalism means practicing at the frontier of knowledge in some field and pushing against that boundary, thus impressing on the students that boundaries need not be (and are rarely) static.

Professionalism for all engineers means an active participation in community life. Engineering cannot achieve general recognition as a profession unless engineers are publicly visible. It is in the realm of public and social service that professionalism shows up strongest. For this reason service to the public and the community and to those less fortunate is particularly significant.

Professionalism can be taught since it is an acquired condition and is not inherent in one's nature. It is most effectively taught by example by individuals whose lives are themselves models of integrity. The beginnings of a professional attitude for the engineering student should be established in the formative college years since, like character, it grows stronger with reinforcement. In laboratory work, for example, an honest reporting of facts and an intelligent evaluation of results are important ingredients in the development of the student's professional training. Design experiences in general involve many compromises—time, money, materials, and so on. Ethical consideration should necessarily become a part of each compromise (decision) made by the young engineer. Your professional career will, in fact, become one of compromise and you should prepare yourself to face the realities of such a life.

Many students will not have achieved a mature professional attitude by the date of their graduation. However, responsibility of thought and decision should be firmly established by this time in order that entry into employment will be a continuation rather than the beginning of professional advancement.

After graduation, opportunities for public service will present themselves. As a part of your professional responsibility, you should seek and accept places of service

Illustration 5-3
Professionalism includes the establishment of performance standards and safety criteria.

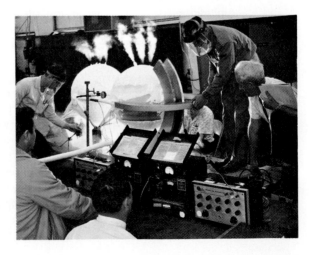

Illustration 5-4

Close attention in class frequently will give major clues pertaining to the nature of future examination questions.

in schools, community government, religious organizations, and charitable groups. Not only will you be able to contribute your talents to these causes, but also you will enhance your own outlook by contacts with both professional and nonprofessional persons. Each engineer should recognize a personal responsibility to acquire a professional attitude and this concern should continue throughout one's professional life.

To sum up professionalism, engineering may be considered to be a profession insofar as it meets these characteristics of a learned professional group:

☐ Knowledge and skill in specialized fields above that of the general public.
☐ A desire for public service and a willingness to share discoveries for the benefit of others.

> Most men believe that it would benefit them if they could get a little from those who *have* more. How much more would it benefit them if they would learn a little from those who *know* more.
> —Wm. J. H. Boetcker

□ Exercise of discretion and judgment.

□ Establishment of a relation of confidence between the engineer and client or the engineer and employer.

□ Acceptance of overall and specific codes of conduct.

□ Formation of professional groups and participation in advancing professional ideals and knowledge.

□ Recognition by law as an identifiable body of knowledge.

With these as objectives, students should pursue their college studies and training in employment so as to meet these characteristics within their full meaning and take their places as professional engineers in our society.

Technical societies

As suggested above, professionals band themselves together for the mutual exchange of ideas, to improve their knowledge, and to learn new skills and techniques. Meeting and discussing problems with others in the same field of endeavor affords an opportunity for the stimulation of thought to improve learning and skills. The National Society of Professional Engineers is concerned primarily with the *professional* aspects of the whole field of engineering. In addition, engineers have organized a number of technical societies in various fields of specialization. For reference purposes, Table 5-1 lists the major engineering and scientific societies in the United States.

Illustration 5-5

Table 5-1

Code	Name	Address	Year organized	Total member-ship
AcSoc	Acoustical Society of America	335 East 45th St. New York, N.Y. 10017	1929	4,100
APCA	Air Pollution Control Association	P.O. Box 2861 Pittsburgh, Pa. 15230	1907	8,100
AAAS	American Association for the Advancement of Science	1515 Massachusetts Ave., N.W. Washington, D.C. 20005	1848	125,000
AAES	American Association of Engineering Societies	345 East 47th St. New York, N.Y. 10017	1980	n.a.
AACE	American Association of Cost Engineers	308 Monongahela Bldg. Morgantown, W.V. 26505	1956	5,000
AAPM	American Association of Physicists in Medicine	335 East 45th St. New York, N.Y. 10017	1958	500
ACI	American Concrete Institute	22400 West 7 Mile Road Detroit, Mich. 48219	1905	14,988
ACM	Association for Computing Machinery	1133 Avenue of the Americas New York, N.Y. 10036	1947	30,000
ACS	American Ceramic Society, Inc.	65 Ceramic Dr. Columbus, Ohio 43214	1898	10,000
ACSM	American Congress on Surveying and Mapping	430 Woodward Bldg. 733 15th St. N.W. Washington, D.C. 20005	1941	6,000
AES	Audio Engineering Society, Inc.	60 East 42nd St., Room 428 New York, N.Y. 10017	1948	4,590
AIAA	American Institute of Aeronautics and Astronautics	1290 Sixth Ave. New York, N.Y. 10019	1932	40,614
AIA	American Institute of Architects	1735 New York Ave., N.W. Washington, D.C. 20006	1857	38,459
AIChE	American Institute of Chemical Engineers	345 East 47th St. New York, N.Y. 10017	1908	36,379
AICE	American Institute of Consulting Engineers	345 East 47th St. New York, N.Y. 10017	1910	420
AIME	American Institute of Mining, Metallurgical, and Petroleum Engineers, Inc.	345 East 47th St. New York, N.Y. 10017	1871	52,000
AIP	American Institute of Physics	335 East 45th St. New York, N.Y. 10017	1931	50,370
AIPE	American Institute of Plant Engineers	3975 Erie Ave. Cincinnati, Ohio 45208	1954	7,794
AMS	American Mathematical Society	P.O. Box 6248 Providence, R.I. 02940	1888	19,400
ANS	American Nuclear Society	555 N. Kensington Ave. LaGrange Park, Ill. 60525	1954	13,000
APS	American Physical Society	335 East 45th St. New York, N.Y. 10017	1899	31,000
APHA	American Public Health Association	1015 15th St., N.W. Washington, D.C. 20005	1872	30,248
ASAE	American Society of Agricultural Engineers	2950 Niles Ave. St. Joseph, Mich. 49085	1907	11,000
ASCE	American Society of Civil Engineers	345 East 47th St. New York, N.Y. 10017	1852	78,000
ASEE	American Society For Engineering Education	Eleven Dupont Circle Washington, D.C. 20036	1893	11,264
ASM	American Society for Metals	Metals Park, Ohio 44073	1913	49,000
ASQC	American Society for Quality Control	161 West Wisconsin Ave. Milwaukee, Wisc. 53203	1946	36,000
ASHRAE	American Society of Heating, Refrigerating and Air-Conditioning Engineers, Inc.	345 East 47th St. New York, N.Y. 10017	1894	40,000

Table 5-1 (continued)

Code	Name	Address	Year organized	Total member-ship
ASLE	American Society of Lubrication Engineers	838 Busse Highway Park Ridge, Ill. 60068	1944	3,470
ASME	American Society of Mechanical Engineers	345 East 47th St. New York, N.Y. 10017	1880	105,097
ASNE	American Society of Naval Engineers, Inc.	1012 14th St. N.W. Suite 807 Washington, D.C. 20005	1888	4,721
ASSE	American Society of Safety Engineers	850 Busse Highway Park Ridge, Ill. 60068	1911	16,000
ASTM	American Society for Testing and Materials	1916 Race St. Philadelphia, Pa. 19103	1898	31,000
AWRA	American Water Resources Association	St. Anthony Falls Hydraulic Laboratory Miss. River at 3rd Ave. S.E. Minneapolis, Minn. 55414	1964	2,400
AWWA	American Water Works Association, Inc.	6666 W. Quincy Ave. Denver, Colo. 80235	1881	25,000
CEC	Consulting Engineers Council of the United States of America	1155 15th St., N.W. Washington, D.C. 20005	1959	2,300
IES	Illuminating Engineering Society	345 East 47th St. New York, N.Y. 10017	1906	10,735
IEEE	Institute of Electrical & Electronics Engineers, Inc.	345 East 47th St. New York, N.Y. 10017	1884	174,000
IES	Institute of Environmental Sciences	940 East Northwest Highway Mt. Prospect, Ill. 60056	1959	1,750
IIE	Institute of Industrial Engineers	25 Technology Park/Atlanta Norcross, Ga. 30092	1948	37,000
ITE	Institute of Traffic Engineers	2029 K Street, N.W., 6th Floor Washington, D.C. 20006	1930	3,800
ISA	Instrument Society of America	67 Alexander Dr. P.O. Box 12277 Research Triangle Park, N.C. 27709	1945	24,736
NACE	National Association of Corrosion Engineers	2400 West Loop South Houston, Tex. 77027	1945	7,425
NAPE	National Association of Power Engineers, Inc.	174 West Adams St., Suite 1411 Chicago, Ill. 60603	1882	12,000
NICE	National Institute of Ceramic Engineers	4055 North High St. Columbus, Ohio 43214	1938	1,625
NSPE	National Society of Professional Engineers	2029 K Street, N.W. Washington, D.C. 20006	1934	80,000
ORSA	Operations Research Society of America	428 East Preston St. Baltimore, Md. 21202	1952	6,800
SESA	Society for Experimental Stress Analysis	21 Bridge Square Westport, Conn. 06880	1943	2,387
SIAM	Society for Industrial and Applied Mathematics	117 S. 17th St. Philadelphia, Pa. 19103	1952	4,990
SAE	Society of Automotive Engineers	400 Commonwealth Dr. Warrendale, Pa. 15086	1905	39,000
SME	Society of Manufacturing Engineers	20501 Ford Rd. Dearborn, Mich. 48128	1932	42,152
SNAME	Society of Naval Architects and Marine Engineers	One World Trade Center Suite 1369 New York, N.Y. 10048	1893	13,000
SPE	Society of Plastics Engineers, Inc.	14 Fairfield Dr., Brookfield Center, Conn., 06805	1941	24,000
SWE	Society of Women Engineers	345 East 47th St. New York, N.Y. 10017	1952	3,000

Problems

5-1. Discuss the factors that are common to all professions.

5-2. Investigate the laws of the state that pertain to serving as an expert engineering witness in court. What would you need to do to qualify as such a witness?

5-3. What engineering fields of specialization are recognized by the state registration board for licensing as professional engineers?

5-4. Locate a copy of an Engineering Code of Ethics and use it as a guide; discuss the procedures that you as a professional engineer in private practice may utilize to attract clients.

5-5. Discuss the value of humanities and social studies courses in relation to the work of the engineer in industry.

5-6. Investigate the need for graduate engineering education for the engineer in private practice.

5-7. List the reasons why it is important for engineers in education to become registered professional engineers.

5-8. Engineer Brown, P.E., is approached by Engineer Smith (nonregistered) who offers a fee of $300 to Brown if she will "check over" a set of engineering plans and affix her professional P.E. seal to them. Describe Brown's responsibilities and actions.

5-9. Discuss the reasons why a professional person such as a registered engineer, an attorney, or a physician will not bid competitively on the performance of a service.

5-10. In interviewing for permanent employment, a senior student in a California engineering school agreed to visit on two successive days a company in Chicago and a company in Detroit. Upon her return home both companies sent her checks to cover her expenses, including round-trip airfare. Discuss the appropriate actions that should have been taken by the student.

5-11. The majority of all engineering designs require some extension of the engineer's repertoire of scientific knowledge and analytical skills. How can the engineer determine whether or not this extension lies beyond the "areas of his or her competence"?

5-12. Engineer Jones, P.E., is the only registered engineer living in Smileyville. Two individuals, Green and Black, approach him with regard to employing his services in estimating the cost of constructing a small dam that would make it possible to reclaim 500 acres of swampland that is now owned by the city but will soon be offered for sale to the highest bidder. Jones learns that both individuals will be bidding against each other to purchase the acreage. Discuss Jones' responsibilities and actions.

5-13. Engineer White is approached by several of his neighbors to urge him to announce his candidacy for city mayor. In previous months the city administration has been accused of "selling rezoning authorizations" and of "enhancing personal fortunes" through the sale of privileged information pertaining to the location of the new proposed freeway. Discuss the course of action that you would recommend for White.

5-14. Engineer Williams and Contractor Smart have been good friends for several years. In March Smart is to begin construction on a multistory building that Williams designed. On Christmas a complete set of children's play equipment (swings, slides, gymnastic bars, etc.) is delivered to Williams' house—compliments of the Smart Construction Company. What course of action do you recommend for Williams?

6

Developing
study habits

From high school
to college

Students who have enrolled in a college or university for the first time often ask, "Is there a difference between a high school course and a college course?" and "Will I need to make any adjustments in my study habits, now that I have enrolled in college?"

The answer to both of these questions is probably *yes,* but let us examine some of the reasons why this may be so.

First, in high school you were competing against the *average* of high school students. However, of the total numbers graduating from high school in the United States each year, fewer than one third go to college. Thus, you are now competing with the average of *very good* high school students.

The study habits and learning process that you used in high school may not be adequate to cope with the increased requirements of college courses because of both the limited time available and the large quantity of material to be covered. A refinement of your study habits or perhaps a complete change in study habits may be necessary to enable you to keep up with the demands of new course material.

Many students, as they enter college, do not realize what will be expected of them. In general they are expected to bring basic skills in mathematical manipulation, in reading rapidly and comprehending, and in possessing a broad-based vocabulary. Engineering educators have observed that a high school graduate who has the ability to *read* and *add* also possesses the capability to succeed in a college engineering program. In high school, much time was taken in class to outline and drill on the daily

> Only the educated are free.
> —Epictetus, ca. A.D. 115

assignments. In college, relatively less time is taken in class, and much more study and preparation is expected from the student outside of class. The student is largely on his own, and his time can be used to a considerable extent as he sees fit. It can be used efficiently and profitably or it can be dissipated without plan and, in effect, be wasted.

Without parental urging or strong encouragement from teachers, the student must adopt personal methods of study that will produce desirable results. Specifically he must budget his time to permit adequate preparation for each course. There must be more than a casual desire to improve study habits. Positive steps must be taken to ensure effective study and learning conditions. It is for this reason that the following topics arc included as suggestions to aid in improving the students effectiveness in study.

Preclass study

The object of study is to learn. Mere idle reading is not study. Particularly in scientific and technical courses, extreme attention to detail is necessary. With the learning process in mind, let us examine some basic principles.

1. The material must be organized into appropriate learning units. Random facts and concepts are more difficult to learn than facts which are related. For example, in learning the names of the bones of the body it is easier to remember the names if groups such as the arm or leg bones are studied as a unit.
2. Attempt to form the correct pattern of facts on the first try. This is necessary to eliminate the need for "unlearning" and relearning factual material. In the case of research or exploratory study, trial-and-error methods are necessary and frequently incorrect assumptions are made. However, by conscious effort to use reasoning and to incorporate other correct facts, false assumptions are minimized.
3. Correct errors immediately and reinforce correct learning responses. Experiments have shown that immediate confirmation of correct learning is more effective in remembering than when the confirmation is delayed. For example, if a mathematics problem is solved and its correctness verified immediately, the principle involved in the solution is retained better than if the verification is delayed.
4. Relate realistic experiences with the facts. Experiments have demonstrated that most people learn and retain information better if it is related in some way to their experiences. For example, an abstract idea such as "democracy" is difficult to present as a realistic picture unless the student has some related background of government upon which to draw a conclusion. On the other hand, a description of a new type of internal combustion engine may be simple to present to an experienced automobile mechanic because of his related experience with similar devices.
5. Give concise meanings to the facts. Particularly in scientific work the meanings of words may not always be clear. Frequently we misunderstand one another because we each may give different meanings to the same word. The use of dictionaries,

encyclopedias, and reference books is therefore necessary to gain a common understanding of new words.

6. Practice, review, and provide application for facts. Education specialists believe that facts are not actually learned until at least one perfect recitation or response is completed. After this has been accomplished, review and repeated use of the facts will greatly aid retention. Research also has shown that if the review is broken into spaced periods, retention and recall are increased (sometimes as much as doubled) over the retention when the reviewing is done all at one sitting. One should be alert to applications for the ideas being learned. This will help to relate them to previous experience and to place them into a pattern where they will become bricks in a wall of knowledge upon which other ideas can be added.

7. Evaluate the adequacy of the learning: A self-evaluation of the understanding of the new ideas which have been presented is one of the most valuable learning experiences in which a student can participate. Memorizing facts does not encourage self-evaluation. However, the ability to apply principles and *to use* facts is one important way in which a person can evaluate the adequacy of learning. For example, after studying a portion of text material in a physics book, are you able at once to apply the facts and principles discussed to the solution of related problems?

The realm of factual information available is so tremendous that a student at first should acquire only the essential and basic facts in a particular field of study. From this set of basic facts, the student then enlarges or details his information into more specialized subjects. For example, the electrical engineering student should begin his study of electricity with a consideration of basic principles, such as Ohm's Law, before beginning to consider the design of amplifiers.

Setting the stage

Provide a designated study area. It is desirable to find a place where you can concentrate and where other people will not bother you. Unfortunately distractions frequently abound in large study areas and interfere with study schedules. Other people in your home or your dormitory may have conflicting schedules and may not be concerned with respecting your own study periods. Radios, televisions, and "bull sessions" are always inviting diversions from study. For this reason many students find that libraries afford good study areas because of the absence of distractions and the ready availability of reference materials. Although a secluded spot is not always essential for study, for most people it does require a conscious effort to reject distracting sounds and backgrounds in order to concentrate.

The best place to study is usually at a desk or table, not on a bed. The effort of sitting helps to keep most people alert and in a mood for study. Good lighting is especially helpful and it is desirable for the whole study area to be illuminated, rather than a small portion of the area. Studies have shown that it is less fatiguing on the eyes if sharply defined regions of light and dark are excluded from the immediate study area. In addition, the work area should be large enough so that reference materials can be kept close at hand.

Prepare a schedule

Time is one of the most important factors to be considered in college study. In every course there is usually more material assigned than can be studied in detail in the time available. In addition outside activities will always compete for a student's time. Athletic events, social and educational programs, recreational activities, and un-scheduled meetings with other people seem to disrupt the best laid plans. The student must realize that these contradictory conditions for study will always exist. Positive steps must be taken to ensure that the time for study is not taken away piecemeal by nonessentials.

In preparing a daily schedule, the question arises concerning the amount of time to allocate for study. Several rule-of-thumb principles are in common use, but the individual's capabilities in learning each specific subject will necessarily need to be the final guide. A recent survey of a cross section of students at a large university showed that the greatest number of students spent an average of 28 to 32 hours of study per week. Engineering students usually spend considerably more than this amount. Actually the number of hours of study is not always the most significant criterion to be used. Rather *how well* one studies is the factor that counts most. The results of study as shown by grades and by one's own personal satisfaction in doing a good job are usually the best indicators of the effective use of time.

A positive and direct approach to a schedule is necessary in the same way that a budget is necessary to manage the fiscal affairs of a business. No commercial enterprise can long exist that does not plan ahead for meeting expenses as they arise. In a similar way a student should prepare a budget of time for his school work and adhere to it, unless circumstances definitely indicate that it should be altered. Not only should the daily time be budgeted on a weekly basis, but also extra time must be allocated for major quizzes, term papers, and final examinations. It has been found, for instance, that the majority of the better students budget their time so that final examinations do not have to be prepared for on a frantic last minute rush. Study

Illustration 6-1
Because practice makes perfect, an unorganized student will in time become an unorganized adult.

skills, no matter how effective, will not be of much value if a student's time is not properly scheduled so that they can be employed.

Scheduling helps to allocate more time to the more difficult courses and to assign less time to the less demanding courses. It also helps to space the available study time so that it will be distributed in a manner to aid in better retention. (Refer to the sixth principle of learning on page 100.)

Studying to learn

Studying to learn is a skill that can be developed just as other skills are developed. All good golfers do not use exactly the same stance or swing, yet they all accomplish a reasonably consistent pattern of results. In like manner, good students may employ slightly different techniques of study and still accomplish an acceptable learning pattern. However, despite individual differences, in any activity a general set of principles can be found that will produce good results, whether it be golf or studying. Some of these principles for study are discussed below.

1. Remove or minimize distractions.
2. Arrange all necessary pencils, reference books, notebooks, note paper, and other supplies before beginning.
3. Put your full attention on the work at hand, and insist that your brain work accurately and rapidly. If the brain is not employed to its full capacity, it will tend to let other thoughts enter to distract it from the task at hand. Read with a purpose—to extract details from the printed page and to comprehend important ideas.
4. Practice reading as fast as feasible. This does not mean skipping from word to word randomly, but rather training the eyes and brain to group words, phrases, and even lines of reading material and to understand the thoughts therein. Many good books are available on how to improve your reading speed for comprehension. Time spent in learning this skill will aid immensely in the faster grasping of ideas.
5. Study as though you were going to teach someone else the subject matter. This will provide motivation for learning and also will encourage self-appraisal of the adequacy of learning as discussed in the seventh principle of learning on page 100.
6. Plan for review and repetition of the assignment. Principle of learning number six on page 100 points out the desirability of spaced periods of review.

The suggested methods and practices described above have been found to aid most students in their learning and the majority of good students follow the general outline of these practices.

Preparation for class recitation

A plan of study for each course is necessary to gain the greatest return in learning from your investment of study time. The plan will vary with the teacher, the textbook, the nature of the course, and the type of recitation and examination that is expected.

> Just as eating contrary to the inclination is injurious to the health, so study without desire spoils the memory, and it retains nothing that it takes in.
> —da Vinci, ca. A.D. 1500

However, before considering suggestions for specific types of courses, we should investigate study techniques that are applicable in general to all subjects.

Learning proceeds best from the general to the specific. It is therefore recommended at the beginning of a course to first skim the chapter and topic headings of the text without reading in detail any of the discussion material. This is done to get an overview of the whole organization of the book. Notice the order in which the topics appear and how the author has arranged the ideas to proceed from one to another. Next, read the lesson quickly to gain an insight into the nature of the material to be covered. Do not attempt to learn details nor to analyze any but the most emphasized points. If there is a summary, read it as part of your lesson survey to prepare you with background material that will be useful in understanding details.

Second, after a rapid survey of the material to be studied, start at the beginning of the assignment with the idea in mind that you will make notes during your study. Remember, you are going to learn as though you would have to teach the lesson content to someone else. Mere superficial reading here will not suffice. The notes can take various forms. They may include summaries of important facts, definitions of words, sample problems or examples, answers to questions, sketches, diagrams, and graphs. A better mental picture is formed and retained if the hand and eye work together on an idea and if you are forced to participate more actively and completely in the learning process.

These notes should be made in semipermanent form, not on random scraps of paper. Bound notebooks or loose-leaf notebooks can be used, but it is important that the notes be organized and retrievable. In addition to separate notes, it is helpful to underline key words and phrases in the text. Don't worry about the appearance of the book. However, do not overdo the underlining; it is better to note the crucial words and phrases so that they are more obvious for review than to underline whole sentences and paragraphs. Usually from three to eight words per paragraph will point out the central idea that has been presented.

Third, reread and review the lesson assignment and prepare your own questions and answers to the topics. At first this may seem to be an unnecessary step but it will pay dividends. Attempt, if possible, to foresee the questions the instructor may ask later concerning the material. Your notes on these predictions can be invaluable at examination time. Check to see that you have noted all the important details and related facts that bring out the main ideas. Particularly in the technological courses, you can do this easily, since much of the material is completely factual.

Fourth, if given an opportunity, plan to participate in the classroom recitation. Force yourself, if necessary, to volunteer to recite. Recitation is a form of learning and it aids in acquiring ideas from others. If the class does not afford an opportunity for recitation, recite the lesson in your room. Review and recitation are the best methods for making a final check of your retention of information. Tests show that you begin to forget even while learning, but if you participate in some form of recitation as soon after study as possible, the retention of facts may be increased by as much as 50 per cent.

Recitation is an effective way of self-appraisal of learning. Just reading a book is not enough to convince anyone—yourself included—that you have learned what you should. When you study, break the topics into groups, and upon completion of each group of topics, as a summary close the book and see if you can recite the important facts either mentally or in writing. When you repeat them satisfactorily then continue; if you cannot repeat them, for further study, go back and pick out the ones that you have missed. It is particularly important to recite if the subject matter consists of somewhat disconnected material such as names, dates, formulas, rules, laws, or items. If the material to be studied is more narrative in style and well organized, the recitation time can occupy a small part of the study period, but it should never be left out altogether.

The general principles above apply to all courses, but certain study plans will apply better to one course than to another. We shall examine study plans for several types of courses in more detail.

Technical courses

In this type of course your study plan should be to direct your study toward understanding the meanings of words and toward grasping the laws and principles involved. In order to understand the words, a dictionary, encyclopedia, and reference books are necessary. The first step is to write down definitions of unfamiliar terms. Remember also that a word does not always have the same meaning in different courses. For example, the word "work" as used in economics has a meaning quite different from the word "work" as used in physics.

When the definitions of words are obtained, study for complete understanding. Texts in technical courses tend to be concise and extremely factual. A technique of reading must be adopted here for reading each word and fitting it into its place in the basic idea. Except for the initial survey reading of the lesson, do not skim rapidly through the explanations, but rather read to locate the particular ideas in each paragraph. If example problems are given, try working them yourself without reference to the author's solution.

After definitions and basic ideas are studied, apply the principles to the solution of

Illustration 6-2
Technical jargon includes not only spoken and written words, but symbols and graphical notation as well. Many lectures will frequently involve the use of all the various media available to the teacher.

problems. It has been said by students that it is impractical to study for examinations where problems are to be solved because you are unable to predict the problem questions. This statement is not correct, for you can predict the principles which will be used in solving the problems. For example, in chemistry a vast number of compounds can be used in equation-balancing problems. However, a very few basic principles are involved. If the principle of balancing is learned, all problems, regardless of the chemical material used, are solved the same way. The objective of this part of study then is to determine the few principles involved and the few problem patterns that can be used. After this, all problems, regardless of their number arrangement and descriptive material, can be classified into one of the problem patterns for which a general method of solution is available. For instance, a problem in physics may involve an electrical circuit in which both current and voltage are known and an unknown resistance is to be determined. Another problem may suggest a circuit containing a certain resistance and with a given current in it. In this case a voltage is to be found. The problems are worded differently, but a general principle involving Ohm's Law applies to each situation. The same problem structure is used in each case. The only difference appears in where the unknown quantity lies in the problem pattern.

Do not become discouraged if you have difficulty in classifying problems. One of the best ways to aid in learning to classify problems is to work an abundance of problems. It is then likely that any examination problem will be similar to a problem that you have solved before.

Learn to analyze each problem in steps. Examine the problem first for any operations that may simplify it. Sometimes a change in units of measure will aid in pointing toward a solution. Try rewriting the problem in a different form. Frequently in mathematical problems, this is a useful approach. Write down each step as the solution proceeds. This approach is particularly helpful if the solution will involve a number of different principles. If a certain approach is not productive, go back and reexamine the application of the principles to the data. For problems which have definite answers, these techniques usually will provide a means for obtaining a solution.

It usually is better in studying technological courses to divide the study periods into several short sessions, rather than one continuous and long study period. For most people a period of incubation (where the idea is allowed to soak into the subconscious) is helpful in grasping the new ideas presented. After returning to do subsequent study, make a quick review of the material previously studied, and look at the notes you have prepared to provide continuity for your thinking.

Literature courses

Most writings classed as literature are written to be interesting and to entertain. For this reason, not as much attention to detail and to individual words is needed as is required in technological books. Usually the ideas are presented descriptively and are readily distinguished. However, the interpretation of the ideas may vary from person to person, and it is with this in mind that the following suggestions are given.

Examine the ideas not only from your point of view but also from the point of view of your teacher. Try to find out from his discussions and examinations the pattern of thought toward which he is directing you, and study the things *in which he is interested.*

Consciously look for these items while you read prose: the setting, central characters (note the realism or symbolism of each), the theme, the point of view of the author (first person, omniscent, etc.), the author's style of writing, the tone, and the type of the writing. For poetry, the ideas may be more obscure, but certain things may be noted. For example, the authors' style, the type of verse, the rhyme scheme, the theme, the symbols, allusions, images, similes, metaphors, personifications, apostrophes, and alliterations are all basic and important parts of the study of poetry.

Social science courses

These courses can be interesting and satisfying or dull and dry, depending upon the student's attitude and interest. Most texts use a narrative style in presenting the material and, as a consequence, the assignment should be surveyed quickly for content and then in more detail for particular ideas. Here the use of notes and underlining is invaluable, and summaries are very helpful in remembering the various facts.

If the course is history, government, sociology, psychology, or a related subject, consider that it contains information that is necessary to help you as a citizen. A knowledge of these subjects will aid you in dealing with other people, and it will give you background information to aid in the evaluation of material that has been specifically designed to influence and control people's thinking. Study the course for basic ideas and information and, unless the instructor indicates otherwise, do not exaggerate the importance of detail and descriptive information.

These principles apply also to courses in economics, statistics, and related courses except that they frequently are treated on a more mathematical basis. Here a combination of techniques described above together with problem-solving procedures can be helpful. Again, since the volume of words usually is quite large, it is necessary to use notes and summaries to keep the ideas in a space to be handled easily.

Language courses

Many techniques have been developed to aid in learning foreign languages. In the absence of specific study guides from your instructor, the following procedures have been found to be helpful.

Learn a vocabulary first. Study new foreign words and form a mental image of them with a conscious effort to think in the new language. As you study, practice putting words together, and, if the course includes conversation, say the words aloud. Space your vocabulary study and review constantly, always trying to picture objects and actions in the language rather than in English.

Rules of grammar are to be learned as any rule or principle: first as statements and then by application. Reading and writing seem to be the best ways of aiding retention of grammar rules. Read a passage repeatedly until it seems natural to see or hear the idea in that form. Write a summary in the language, preferably in a form that will employ the rules of grammar which you are studying. Unfortunately, there is no way to learn a new language without considerable effort on your part. Even English, our native tongue, when studied as a subject, gives some students trouble. However, many students have said that they really understand basic English much better after having taken a foreign language.

Classroom learning

The discussion so far has been concerned with learning by study. An equally effective and more widely used method of learning is by listening. From earliest childhood you have learned by listening and imitating. Do not stop now but rather use the classroom to supplement your home study. You will find that things are covered in classroom work that you do not find in your texts. The interchange of ideas with others stimulates your thinking and retention processes. The classroom can also be a place to practice and to demonstrate your learning and problem-solving skills before the examination periods.

The skill of listening seldom is used to the fullest extent: If your attention is only partly on the lecture, the part missed may make a major difference in your grade. Use the time in class to evaluate your instructor, find out what he will expect of you, watch for clues for examination questions, and make notes to be used for later study. If you plan to make the classroom time profitable, you will find it also will be enjoyable.

Come to class with a knowledge of the assignment to be discussed. The instructor then can fill in your knowledge pattern rather than present entirely new material. This also saves time in taking notes because the notes will be needed only for amplification rather than as semiverbatim recording. If a point arises at variance with your knowledge from study, you have an opportunity to question it. Prior study also permits you to predict what the instructor will say next. This serves as a valuable psychological device to hold your attention throughout the class period.

Note taking during class is a skill that can be learned. The inexperienced will try to take notes verbatim and thus get so involved in writing that they cannot listen for ideas. Usually, they cannot write fast enough to copy all the words anyway. Rather than take notes verbatim, practice your listening skills and evaluate the critical points in the lecture. A few critical points will be amplified with descriptive materials. Practice taking down these critical points in your own words. Such note taking serves to keep your attention focused and to encourage a better understanding of the principles being discussed. If you do not understand the points completely, make notes for later study or for questioning the instructor.

If the course is such that you can, recite during the class period. Push your shyness aside, and place your ideas and information before the class. It stimulates your thinking and retention, it will help to clear up obscure points, and will give you much needed practice in hearing your voice in the presence of others.

> I am a great believer in luck. The harder I work the more of it I seem to have.
> —Coleman Cox

Attention and listening during a class period together with participation, either in recitation or in anticipating what the instructor will cover next, will save you hours of study time outside the classroom.

Preparation for
themes, papers, and reports

The purpose of writing is to transmit information. For an engineer this is a valuable means of communicating his ideas to others and the practicing engineer takes pride in the conciseness and adequacy of his reports. No matter how good your ideas may be, if you cannot communicate them to others, they are of little value. Since part of the work of an engineer is writing reports and papers, the opportunity to learn and to practice this skill in school should be exploited. There are many good books on composition and manuals on writing that will help you. For this reason the suggestions given here are to be considered as supplementary helps in the preparation of written work.

In general, good writing involves good grammar, correct spelling, and an orderly organization of ideas. The basic rules of grammar should be followed and a logical system of punctuation used. If in doubt as to the application of a somewhat obscure rule in grammar, either look it up or reword your idea in a more conventional manner. Punctuation is used to separate ideas and should follow, in general, the pauses you would use if you were reading the material aloud. Usually, reports are written in an impersonal manner; seldom is the first person used in formal writing.

Little needs to be said about spelling except to say, spell correctly. There is so little room for choice in spelling that there is no excuse for a technical student to misspell words. If you don't know how to spell a word, look it up in a dictionary or reference book and remember how to spell it correctly thereafter.

The last characteristic of good writing is a clear orderly organization of ideas. They may take several forms depending upon the type of writing. For themes, usually a narrative or story form is used in which a situation is set up, possibly with characters, and a story is told or a condition is described. Engineering papers and reports generally are concerned with technical subjects. Therefore, they describe the behavior of objects or processes, or they provide details of events in technical fields. Frequently, the first paragraph summarizes the thoughts in the whole report in order to give the reader a quick survey without his having to skim through the manuscript first. Following paragraphs outline the contents in more detail. They frequently end with graphs, drawings, charts, and diagrams to support the conclusions reached.

In preparation of written work, some research usually is needed. In order to aid in keeping the notes for the material in usable form, it is helpful to record abbreviated notes from research works on cards in order that the arrangement of the writing of the paper can be made in a logical order. Usually, in compiling information you do not know how much will be used so the notes on cards provide a flexibility of choice that is a great aid in the final organization of the paper. The cards can be 3 by 5 in. or 4 by 6 in., with the latter usually being the better choice because of more available space.

An outline of the material to be discussed or described is necessary, even for brief reports. An outline ensures a more logical arrangement of ideas and helps to make the writing follow from concept to concept more smoothly.

Write a draft copy first and plan on making alterations. Write first to get your ideas down on paper, and then go over the copy to improve the rough places in grammar, spelling, punctuation, and wording. If possible, wait a short time before taking these corrective steps to get a more detached and objective approach to the suggested changes. When the rework of the draft has been made, copy it over neatly, still maintaining a critical attitude on the mechanics of the writing. It is helpful if the final draft can be typed, but if not, you will find that good handwriting or lettering frequently makes a favorable impression upon the person who grades your paper.

Preparation for examinations

Have you ever felt after taking an examination (for which you studied) that everything you studied was inappropriate? Perhaps you used the wrong techniques in studying for the examination. Certain rules have been found to be very useful in preparation for tests of any kind. The type of preparation you make is often more important in the final grading than how long you spend in preparation. Let us discuss some of these rules that have been found to be effective.

Start preparing for tests the day the course starts. You know that they will be assigned, so do not close your eyes to this fact. From the very first, start studying two things: (1) the big overall ideas of the course, and (2) the instructor. Keeping these two things in mind will help you to learn while in class and while studying. This will also mean shorter reviews before tests.

Illustration 6-3
In preparing for tests, it is frequently advisable to acquire a study partner.

Keep writing reminders of important points—as notes in the margins of your books, as flagged notes in class, or as short statements while studying. Use nontext material as it becomes available, such as outlines, old tests, and information from students who have taken the course.

It is not unethical to study the instructor. After all, he is a person qualified to present the subject matter, and, because of his training and experience, frequently you will learn much more from him than you will from any text. From a study of the instructor, you can follow his pattern of lesson organization and find out what he wants from you. In your preparation, attempt to think and study along these lines. Close attention in class frequently will give major clues pertaining to the nature of future examination questions.

Make a final review of the subject matter. This should be a planned review and not a "last ditch" cramming session. Schedule it in several short sessions and do these things: First, review the general organization of the material before the final class periods of the semester in order to take advantage of the instructor's summaries and reviews. Second, set up the major topics or ideas and associate them with specific facts or examples. In the case of problem courses, this is the time to work out and review sample problems that will illustrate laws or principles. Third, study for more detailed information and to complete the areas of uncertainty. If the first two phases of the review have been adequate, this last phase should take relatively little time.

Predict the type of test that will be given. If the test is to be an essay type, practice outlining key subject matter, summarizing important concepts, comparing or contrasting trends, and listing factual data. This may seem to be an excessive amount of work, but if you have noted the points the instructor has stressed during the course, you can narrow the field considerably. However, a word of caution here: Do not try to outguess the instructor; it usually will not pay. Study the topics you honestly feel are important, and avoid unjustified evaluation of different concepts in hope that there will not be questions on them. Remember that on this type of test not only are the ideas to be recalled but also the organization and sequence of the ideas is important.

For objective or short-answer type tests, follow the three-step program of study given above, giving more attention to relating key ideas to specific items of information. Here, short periods of highly concentrated study usually are to be preferred. Think about each idea long enough to form mental pictures and precise answers, but guard against merely memorizing words. Frequently sketches and diagrams will aid in retaining a mental picture of a concept.

If the test is to be a problem-solving test, first review the principles that may be encountered in a problem. Work at least one sample problem that will illustrate the principle. Ask yourself, "If there is a change in the quantity to be solved for, can I still place this problem in the correct problem-solving pattern?" This is important because, although the variation in problem statements is infinite, the applicable principles and consequently the problem patterns are relatively few.

What if you have more than one examination on the same day? Most students have found it better to do the last review on the subject that comes first. A quick check of notes then can be made before the next examination begins. An old but useful maxim states that the best preparation for taking tests is to practice the things you will need to do on the test.

Taking examinations

We shall assume that a primary objective in taking an examination is to make a high grade on it. Your grade will be based on what you put down on the test paper—not on what you know. It is crucially important then to get the correct sampling of your knowledge on record. Sometimes students fail, not because of lack of knowledge, but rather because of lack of skill in proving on the test paper that they do understand the material.

In taking any examination, be prepared. The ability to think clearly is of most importance, but the ability to recall facts is also very important. Enter the examination room with a feeling of confidence that you have mastered the subject and, while waiting for the questions to be distributed, be formulating plans for taking the test so that your mind will not be blocking itself with worry.

If the test is an objective type, turn quickly through the pages and note the kinds of questions; true–false, multiple choice, matching, completion. Make a rapid budget of time, and read the directions for answering the questions. If there is no penalty for guessing, answer every question; otherwise plan to omit answers to questions on which you are not reasonably sure. Be certain you understand the ground rules for marking and scoring so that you will not lose points on technicalities.

A basic principle of taking any examination is: *Answer the easy questions first.* If time runs out, at least you have had an opportunity to consider the questions you could answer readily; and usually an answer to an easy question counts as much as an answer to a difficult question. Do not carry over thoughts from one question to another. Concentrate on one question at a time and do not worry about a previous answer until you return to it on the next trial. It helps to relax for a moment between questions and get a fresh breath, and to help dismiss one set of thoughts before concentrating on another. Look for key words that may point to whether a statement is true or false. Usually statements are worded so that a key word or phrase tips the balance one way or the other. In case of doubt, try substituting a similar word into the statement and see whether it may aid in identification of truth or falsity. When you go back over the examination, do not change your answers unless you have obviously misread the questions or you are reasonably certain your original answer is incorrect. Tests have shown that your first response to a question on which you have some doubt is more likely to be correct than not.

If the test is an essay type, again read quickly through the questions, budget your time, and answer the easy questions first. It is helpful to plan to put an answer to each question on a separate sheet of paper unless the answer obviously will be short. The one-answer-per-page system permits easy addition of material after your initial trial. Watch your time schedule, since it is easy to write so much on one question that you are forced to slight others. A help on answering lengthy questions is to jot down a hasty outline of points to include so they will not be overlooked in the process of composition.

After the questions have been answered, take a final critical look at your paper to correct misspelling, grammatical errors, punctuation, and indistinct writing. If time permits, add sketches, examples, or diagrams that may come to mind. Sometimes a period of quiet contemplation, mentally reviewing your notes, will help recall needed additional facts.

If the test has mathematical problems to solve, again read through the questions and budget your time. Plan to answer the easy questions first. Determine the "ground rules," such as whether points will be given for correct procedure regardless of the correctness of the arithmetic. Unless the problem solution is obviously short, plan to work only one problem per page. If a mistake is detected you can more easily and more quickly line out the mistake than attempt erasure. One answer per page also permits room for computations and makes checking your work easier. Usually it is better to do all the work on that page and avoid scratch paper.

If the test is an open book test, use the reference books only for tabular or formula data that you reasonably could not remember. If you try to look up things you should already know, you will surely run out of time.

Let each problem stand by itself. First, analyze it from the standpoint of a pattern into which it can be fitted. Consider then what steps will need to be used in the solution, and finally determine how these solution steps will be presented. When the analysis is complete, solve the problem in the framework of the analysis.

Usually it is better to go ahead and work through all problems and then come back and check for arithmetic mistakes and incorrect algebraic signs. This is the place also to take an objective look at the answer and ask whether it seems reasonable. A questioning attitude here may reveal mistakes that can be corrected.

Analysis of results of tests

Finally, when the test is ended and you get an opportunity to see your graded paper, analyze it and yourself critically. Assuming that you knew the material but that your grade did not reflect your knowledge, find out why the grade was not as good as it should have been. Blame only yourself for any deficiencies. Look for clues such as the ones given below that will help you not to make the same mistake again.

If it was an essay test, was your trouble poor handwriting, incorrect grammar, incorrect spelling, or incorrect punctuation? Correction of these faults is a matter of the mechanics of learning the rules and making a conscious effort to improve on your shortcomings.

Was your trouble failure to follow instructions, lack of organization of ideas, or lack of examples? Look for clues such as marks on your paper by graders stating "not clear," "not in sequence," "why?," "explain," "?," "trace," "compare," "contrast," and so forth, which indicate a failure on your part to follow instructions. The remedies are twofold. Look for key words in instructions on tests, and practice beforehand the listing, contrasting, or comparing of factual data.

When the grader's marks include words such as "incomplete," "hard to follow," "meaning not clear," or "rambling," these comments indicate that your ideas need to have better organization. A remedy is to consider carefully what is being asked for in the question. Make a brief outline before you start writing. This affords a means of placing ideas in the most effective sequence and also helps to avoid omitting good points.

The grader's marks may be "for example," "explain," "be more specific," or "illustrate." These marks usually indicate a need for illustrations and examples. Your answers may show that you know something about the subject, but they may not convey precise information. Examples will convince the grader that you know the material covered.

For objective tests, evaluate the patterns of the questions missed. Did you misread the questions? Were you tripped up by double negatives in true–false questions? Did you fail to look for key words in multiple-choice questions? Did you realize immediately after the examination that you had answered incorrectly? Some aids in improving grades on objective tests follow.

For true–false questions, did you give each question undivided attention, and were you careful not to read something into the question that was not there. If you missed several questions in sequence, you probably were thinking about more than the question at hand. Try rewording questions that have double negatives next time if you show a pattern of missing them.

For multiple-choice questions, determine whether you concentrated on each question alone and determined, if possible, what the answer was before looking at the set of multiple-choice answers. You should have eliminated as quickly as possible answers that obviously did not fit the question and concentrated on key words that would have provided clues to select from the remainder.

If the test consisted of problems to be solved, check for mistakes in two things: analysis and arithmetic. If your paper shows false starts on a problem, if you worked partway and could go no further, or if the solution process was incorrect from the beginning, your principal trouble probably is lack of skill in problem analysis. The remedy, of course, is to work more problems illustrating the principles so that the test situations will be more familiar. If you use no scratch paper on tests, and keep all parts of your solution on the page, checking to ascertain your mistakes should be easy.

If the processes are correct but the answers are incorrect, look for careless mistakes: in arithmetic, in employing algebraic signs, in mixing systems of units such as feet and centimeters, in copying the problem or in copying from one step to the next, or in making numbers so indistinct that they are misread. The remedy for these mistakes is to go over the solution carefully checking for these things. If time permits, one independent solution will help. Finally, look at the answer—does it seem reasonable?

The employment of the techniques discussed above should help you to achieve grades based on your knowledge of a subject without a handicap in the skill of presenting the knowledge on an examination. No more should you have to say, "I knew it but I couldn't put it down on paper."

7
Spoken and written communication

Compared to Europeans and even to other nationalities who speak English, we Americans have a peculiar uneasiness about our language. Some of us are reluctant to speak out because we are shy about our "grammar." Others confuse language ability with the ability to spell. Many Americans think English is difficult in some obscure way. It is a common misconception that English is extraordinarily difficult for foreigners to learn.

Some linguists attribute our attitude toward language to the American self-image: we think of ourselves as men of action, tight-lipped, monosyllabic, like Gary Cooper with his "yup" and "nope." But what really sets us apart from much of the world is that we have no language establishment. France has a venerated Academy that sets standards, Great Britain a university-bred establishment, and in the totalitarian states the ruling party controls the language along with education. In contrast, we have no "official" American English.

To further add to our confusion, linguistic researchers have revealed some new discoveries about language that are going to demand some rethinking on our part. Just as today's anthropologists are learning that some societies of the past that have been shrugged off as "primitive" were in many respects just as complex as ours, linguists are finding that no language is really primitive. For example, the Australian Aborigine speaks with a grammar every bit as complicated as ours. Every society seems to develop a language that wholly satisfies its need.

In some cultures, linguistic skill is lavishly rewarded; in others it is regarded with suspicion. The American attitude lies somewhere between these extremes. In general, we admire the plainspoken man, but also chuckle at Li'l Abner's crude English. A generation ago, we responded to the power of Winston Churchill's oratory, but today anyone who oversteps such very narrow bounds might well be labeled a windbag politician.

Skills in communication are important for the engineering student and for the

> A bolt is a thing like a stick of hard metal such as iron with a square bunch on one end and a lot of scratching wound around the other end.
>
> A nut is similar to a bolt only just the opposite being a hole in a little chunk of iron sawed off short with wrinkles around the inside of the hole.
>
> —An 8th-grade student's definition and description of two mechanical objects.

engineer in practice. If an engineer cannot express clearly his ideas and the results of his endeavor to others, even though he may have the intellect of a genius and the capability of performing the most creative work, the benefits of his intellect and creative abilities will be of little use to others. A surprisingly large amount of an engineer's time and effort is devoted to communicating—principally writing and speaking. While in college, most engineering students don't recognize this, and they are not easily convinced that the ability to write and speak effectively has a *significant* effect on their professional competence.

What are the skills that are needed in communications? For the engineer they generally are classed as verbal, graphical, and mathematical. In this respect we shall consider that *verbal* means language communication, either oral or written; that *graphical* constitutes all pictorial language such as engineering drawings, charts, diagrams, graphs, and pictures; and that *mathematical* includes all symbolic language in which concepts and logic processes are presented by use of a system of prearranged symbols.

A question may arise as to whether models and demonstrations constitute communication. In the truest sense, they do, but usually they are inadequate within themselves to convey all concepts. Since usually words, pictures, or symbols are used as a supplement to explain such devices, these methods should not be considered to be a separate means of communication.

Illustration 7-1

The engineer must be able to convey ideas to others by means of spoken and written communications. Conciseness without loss of clarity is most desirable.

Verbal communication

We begin practicing oral communication at an early age. Unfortunately, the basics of English composition, spelling, and grammar and an understanding of what the language is and how it works *are not* being effectively mastered in the public schools by an increasingly greater percentage of students. For this reason, the student who realizes the need and has the desire to improve his communication skills should either avail himself of college courses in speech and English composition or undertake a self-study program.

As we progress through school, we add to our vocabulary and pick up experience in presenting ideas in writing. By the time a student is a freshman in college, he is expected to have a working vocabulary of several thousand words, to be able to organize ideas into a coherent pattern, and to present these ideas either orally or in writing. How does the objective of verbal communication for an engineer differ from this objective for other people?

Primarily, the engineer communicates to present ideas and to gain ideas. For some people, talking or writing is purely an entertainment, or an outlet for creative feelings. On the other hand, for the engineer, verbal communication is a part of his professional life. He communicates on a technological level with other engineers and with scientists and technicians and on a layman's level with nontechnically trained people.

Although public awareness and understanding of engineering, science, and technology are increasing each year—the world is becoming more technologically complex—many people are uneasy or uncomfortable about "what technology is doing." In their minds, the "what" is frequently assumed to be bad or, at least, constitutes a veiled threat because of its mysteriousness. The mystery, of course, is a result of their lack of understanding and knowledge in these areas (the United States' conversion to the metric system is too often met with fear, or at least discomfort, by large numbers of people).

The point of all this for the engineer, of course, is that he has an important job to do in selling his competence, knowledge, and judgment. His profession and its activities must be understood and appreciated by the public—not condemned through ignorance. In general, his works bring about much public good, but he must let this be known broadly.

His skill in communicating to the nontechnically trained public is not only a professional responsibility, it is his personal responsibility. Ideally, an engineer should be able to take a most complex topic—nuclear fission, a jet turbine, holography, or systems analysis—and explain it in such a way that his brother in the fifth grade or his Aunt Bessie will have a basic, simple understanding of it.

For example, suppose you have been working on a device for inclusion in the design of an autopilot control for an airplane. This device includes a potentiometer and a gyroscope. Consider how different your description of the device would be to an engineer and to an accountant. The ability to "speak their language" is an important skill which the engineer should possess in dealing with diverse people.

Not only should the engineer's verbal skill be descriptive, but also it should be persuasive. Frequently, good ideas are considered by the uninformed to be too impractical or too revolutionary. The engineer should have not only an adequate vocabulary, but also skills in presenting his ideas in a way that others will be led to accept them. In situations like this, practice in idea organization, a knowledge of psychology, and training in debate are all helpful.

In college, many opportunities are available for participation in group discussions, for the presentation of concepts both written and oral, and for gaining vocabulary skills. This is the time for the engineering student to learn by trial and error his best ways to communicate. After graduation, trial-and-error methods may be economically impractical. A conscious effort while in college to improve one's ability to communicate verbally will make the transition to work as a practicing engineer after graduation much easier.

Graphical communication

How often have you heard someone exclaim, after a futile attempt to describe an object to another person, "Here, let me draw you a picture of it!" The old adage of a picture being worth a thousand words is still true. The ability to present ideas by such means as pictures, diagrams, and charts is a valuable asset. In general, the engineer is expected by nontechnical people to be able to sketch and draw better than the average person. Today, most engineering curricula include some work in engineering graphics, and although the engineer may not be a professional draftsman, he should be able to attain and maintain an acceptable level of performance in engineering drawing and lettering.

Since ideas in research and development are frequently somewhat abstract, diagrams and graphs not only help to present ideas to others but also help the engineer to crystalize his own thought processes.

For the design engineer, the ability to present ideas graphically is a necessity. In almost every case, instructions prepared by a design engineer for use by technicians or workmen in building or fabricating articles are transmitted in the form of drawings. In the case of machined parts, for example, usually the workman has only

Illustration 7-2
The engineer must be able to understand graphical as well as oral and written instructions.

the vaguest idea of the application of the part, so the drawing prepared by the design engineer must tell the complete story to the machinist.

The engineer engaged in sales will need to make frequent use of graphic aids. It usually is easier and faster to project ideas and application by graphic means than by verbal communication alone. Pictures fill the gap between verbal description and actual observance of an operating device. So effective are these techniques that considerable experimentation is now being done in teaching by means of television.

Mathematical communication

Mathematics involves the use of symbols to represent concepts and their manipulation in logic processes. It has been stated humorously, but nevertheless somewhat truthfully, that mathematics is a form of shorthand used to describe science and that higher mathematics is shorter shorthand. In his study of mathematics, the engineer learns the meaning of symbols such as π, $+$, and $\int$, the rules for manipulating mathematical quantities, and the logic processes involved.

The question sometimes arises as to why an engineer needs so much mathematical training. The answer in simple terms is that it is such a valuable and powerful tool that the engineer cannot afford to ignore its use. By using mathematics, not only is space conserved in the presentation of ideas but also the task of carrying the ideas through logic processes is simplified. Since many engineering science operations follow elementary mathematical laws, it is much easier to transform ideas into symbols, and manipulate the symbols according to prearranged mathematical procedures, and finally to come up with a set of symbols which can be reconverted into ideas.

The engineer's way of thinking is so consistently geared to mathematical processes that it becomes almost impossible for him to think otherwise. For example, if you are asked to find the area of a circle whose radius is known, you may immediately visualize $A = \pi r^2$. Now try to think of finding the area of a circle without using such a mathematical formula—you will probably find such thought to be difficult and unnatural.

Since your mathematical training has given you a skill in communication, as an engineer you should make full use of it. As has been pointed out, the logic processes enable one to predict mathematically the behavior of many engineering science operations. In addition, the mathematical presentation of the ideas enables others familiar with mathematical rules to envision the practical application of the concepts.

For example, if the effects of gravitational forces, air loads, centrifugal forces, temperatures, and humidity are expressed properly in a mathematical formulation, the path of a missile over the earth's surface can be predicted with surprising accuracy. It is not actually necessary to perform the flight and to measure the trajectory if the parameters involved are known accurately.

Of course, mathematics is not restricted to an application of the known behavior of objects. By mathematical extrapolation, fundamentals of natural laws have been determined even before it has become known that such behavior is possible. For example, the principles of atomic fission were predicted mathematically many years before it was possible to verify them experimentally.

The use of mathematics by engineers permits more time to be given to creative thought, since ideas can be explored symbolically without having to make physical determinations. Of course, the advent of high-speed automatic computing machines also has aided both in accelerating exploratory research and in executing routine mathematical operations.

Technical reporting

One of the first things a young graduate must learn is that writing or speaking as an engineer on the job is not at all like writing or speaking as a student in college. For example, engineering students usually write for professors who know more about the subject than they do. On the job, however, the engineering graduate will write for management, for technicians, for skilled and semiskilled workers, and only occasionally for other engineers. Therefore, in college the young engineer writes to impress the reader with his knowledge and diligence; on the job, he writes to inform, to instruct, and not infrequently to persuade. However, for both audiences his goals are clarity, logic, and economy of words.

Much of the engineer's communication is executed by reports. These reports may be oral presentations in the form of technical talks or they may be written presentations as technical reports. In either case, information must be presented in a form so that the desired meaning can be understood.

Since the objective of a report is to present information, it must be prepared with the reader in mind. Clarity is therefore a prerequisite for a good report. A report that uses rare words or uncommon foreign phrases may serve to point up the brilliance of the author, but it may also discourage readers from attempting to unravel the

Illustration 7-3
Technical talks should be supplemented with appropriate charts, slides, graphs, models, and so on, to add clarity and interest.

meaning. A report should be prepared using words and phrases with which the reader will be familiar.

In addition to clarity, a report should state clearly and honestly the results obtained. In the case of reporting on tests, frequently data are taken and the test assembly is dismantled before the test results are available. Therefore, the tests cannot always be rerun, and the data are usually used as recorded. If the results should turn out to be less than desirable, as an engineer you are obligated to report the facts completely and honestly. Even though reporting the true facts may be distressing to the writer, an honest statement will instill a feeling of confidence in the reader that the results are trustworthy.

In preparing reports, in general, only factual material should be covered. There is often a temptation to include irrelevant subjects or personal opinions as a part of the factual material. In some cases, it may be desirable to give a personal opinion, but such opinion should be identified clearly as a matter of judgment and not as factual data.

Written reports

There may have been a time when engineers were not required to write reports, that is, when most of them worked with small groups of people and they could let their ideas be known by word of mouth or by circulating an occasional sketch or drawing. These times are gone for all but a very few engineers. Most engineers today work in large organizations. They cannot be "heard" unless their ideas are written down in proposals and their findings are recorded in reports. This does not mean that the spoken word and the drawing or sketch have lost their importance, but rather that they must be supplemented by the written word. Therefore, it is important for you to know how best to communicate your ideas to a reader; how to put your best foot forward with a client whom you may never see, or with the company vice-president to whom you will report.

As an engineer, your writing is often directed toward other engineers, occasionally may be called upon to write for an audience unfamiliar with technical terms. Then you must be able to express your thoughts in terminology that can be understood by an intelligent layman. During the years to come, when engineers must solve the problems of society, such as the urban crisis, air and water pollution, energy, food production, and so on, the need for cooperation between technical and nontechnical persons becomes increasingly important. The engineer must learn to communicate with people of all types of background, and must be able to state views clearly and concisely.

In writing, all engineers need to be aware of a few common sense rules of technical communication. They should not try to show off their knowledge, but should

> Short words are best and the old words, when short, are best of all.
> —Sir Winston Spencer Churchill

keep their vocabulary as simple as possible without "talking down" to the reader. It should be kept in mind that the reader may not be a specialist (most engineers direct their writing to management) and that this same reader is probably in a hurry. He or she does not want to be held in a prolonged state of suspense awaiting some thrilling outcome. Rather, what the reader wants is information as quickly and as painlessly as possible, and expressed in a way to connect it in some manner with his or her own experience.

Keeping the reader in mind, the successful engineer/writer will organize required reports in reverse order to the way that the work is accomplished. An engineer proceeds in design work from problem recognition to solution, bringing order out of chaos, isolating bits of data, selecting and interpreting, and, after trial and error, arriving at a specific finding. This finding might be a particular choice of structural material, an optimal shape, or even the decision to terminate a project.

When your design or analysis work is complete you are ready to write your report—but in reverse order. You must avoid the temptation "to write a detective story"—the temptation to lead the reader from the beginning over the same arduous route that you yourself followed in your quest for truth. (All detective fiction is synthetically organized this way.) Instead, you should organize your paper to reveal

> Anyone who wishes to become a good writer should endeavor, before he allows himself to be tempted by the more showy qualities, to be direct, simple, brief, vigorous, and lucid.
> —H. W. Fowler

all major conclusions in the first few paragraphs. Now the reader immediately knows what the paper is about. If you choose, you may read further to find out why your conclusions are true and *how* you arrived at your decisions.

So many books have been written about the art of writing, about grammar, syntax, and style, that it would be presumptuous to try here to summarize them in a few words. However, we would like to quote a few phrases from a small but exceedingly valuable book on style.[1] These authors recommend some 21 rules, among which are the following:

☐ Place yourself in the background. Write in a way that draws the reader's attention to the sense and substance of the writing.

Illustration 7-4
Accuracy, brevity, and clarity are fundamental to good communication.

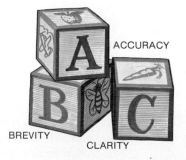

ACCURACY

BREVITY

CLARITY

[1] William Strunk, Jr., and E. B. White, *The Elements of Style,* third edition (New York: Macmillan, 1979).

- Write in a way that comes naturally. But do not assume that because you have acted naturally, your product is without flaws.
- Write with nouns and verbs, not with adjectives and adverbs.
- Revise and rewrite.
- Do not overstate, because it causes your reader to lose confidence in your judgment.

If we can assume that we know how to write, how to express our thoughts in words and sentences that are clear to the reader, we still need to know how to organize our ideas. Organization is important to the writer so that his ideas will be presented in a logical sequence and to assure that the important things are included in his writing. Organization is important to the reader so that he can follow the presentation and conclusions of the author easily, so that he need not jump back and forth in his thoughts (a tiring exercise for any reader).

Over the years certain minimum conventions (standards) have been established concerning the writing of engineering reports and proposals, conventions that are not binding but have proven to be useful guidelines for technical writers. Let us look at the typical organization of such a report.

Organization of a technical report

Abstract
Table of Contents
Table of Figures
Acknowledgments
Nomenclature

Introduction

Body of report
Analysis
Design
Experiments
Test Results

Discussion

Summary and Conclusions
References
Appendixes

The essential features of the technical report are shown in italics. The other items *may* appear in the report if appropriate.

The *introduction* tells what the problem is and why it was studied. It will discuss the *background* for the study, the literature that pertains to the subject, the solutions that have been tried before, and why these are not adequate for the present investigation. It is here that the majority of the literature references are mentioned. If there are three or less, it may be adequate to list them in footnotes. However, when there are more than three references, it is common practice to list them together in a reference section at the end of the report.

The *body of the report* may have any of a number of titles and may, in fact, consist of several chapters with different titles. The author has considerable latitude here and should make use of titles that appear to be appropriate. For example, if the work was essentially analytical in nature, it may be appropriate to entitle the section "Analysis," or perhaps to be more specific and to discuss first the assumptions that were made, then the construction of the model, the pertinent equations, and finally the solution of the equations. If the report contains information on experiments, the writer may wish to discuss the experimental apparatus, the construction of the test model, and the peformance and organization of the test. This may then be followed with a chapter discussing the test results.

The preparation of the body of the report requires considerable judgment. The engineer must provide enough information to give the reader a very clear picture of what was done and to allow him to arrive at the *conclusions* of the report. On the other hand, it is essential that the reader not be bored by unnecessary detail. Many writers find it appropriate to give only the major outline of their work in the body of the report and to relegate all important but minor details to appendixes at the end of the report. This technique gives a report a highly desirable conciseness.

It is usually expedient to illustrate the body of the report with tables, charts, graphs, sketches, drawings, and photographs. The old adage that "one picture is worth a thousand words" is often true, but care should be taken to avoid unnecessary illustrations.

Another important feature of functional technical writing is the use of headings and subheadings. These allow the reader to skip over parts already known, or parts that are too specialized to follow, or parts that simply are of no interest.

Since it is customary to limit the body of the report to facts, the *discussion* section permits a review of the author's opinion. It is as if one could stand back and look at the work and say why this or that was done, to speculate on why the results are the way they are, and what they might have been if the experiment had been done differently. The discussion sections should anticipate the type of questions the listener would ask and attempt to answer them as forthrightly and honestly as possible.

The *summary and conclusions* is, as the name implies, a concise statement of the work done—including goals, background, analysis, experiment, and a review of the work accomplished. The concise statement of the conclusions reached is most important. For the reader, the conclusions should be the "pot of gold" at the end of the rainbow, the information that will be of direct use. Therefore, the development of meaningful conclusions, well stated, is one of the most important parts in writing a report. They should include all that is new and important, and yet they should be so stated that they leave no question in the reader's mind as to what is incontrovertible fact and what is opinion. Wherever possible the writer should make estimates of the accuracy and repeatability of his results. It is often useful to number the conclusions much as a patent attorney will number the claims in a patent application.

After the conclusions have been written, the author should write the abstract as if it were a summary of the "summary and conclusions" just finished. Only the most important conclusions need be included in the abstract.

A note on the convention for *references*. In most engineering reports, it is now customary to list the last name of the senior author first, followed by his initials, and followed by initials and name of coauthors. The names are then followed by the title of the report and this by the name of the journal in which it was published, or the publisher and year, in case it is a book. Typical references are as follows:

Smith, A. B., and T. D. Jones, "Air Pollution at the North Pole," *J. Arctic Society,* Vol. 15, No. 6 (1964), pp. 317–320.
Beakley, G. C., *Introduction to Technical Illustration,* Indianapolis, Bobbs-Merrill Educational Publishing, 1983.

The formal report of a feasibility study often may be in the form of a proposal which suggests how the problem should be pursued. There are many similarities between an engineering report and a proposal, but their purposes are quite different. The report exists to present the results of a study and to present them so clearly and completely that other engineers can use them as stepping stones in the further development of engineering knowledge and use. The proposal, on the other hand, proposes to sell an idea—tries to convince a client or a superior to make funds available for the preliminary design. Thus, while the report is written for a general audience, not necessarily all engineers, the proposal is always written for just one person or organization. A proposal is an attempt to sell an idea. Therefore, what is good advice for the salesman is also good advice for the proposal writer: *try to put yourself in the position of the client.* Find out what his needs and wants are and see to what extent your idea meets these needs. Find out who else competes for the funds which might be used to further your idea and emphasize those special points that make your idea or talents superior to that of others.

There is no general format for the organization of a proposal, but the following order is used frequently.

Typical organization of a technical proposal

Technical Part
Introduction
Objectives
Background
Method of Approach
Qualifications
Management Part
Statement of Work
Schedule and Reporting
Cost Estimate
[Other special paragraphs, i.e.,
   Rights to inventions
   Security provisions
   Time at which work can begin
   Time limit on proposal acceptance]

The proposal is often split into a technical part which discusses the technical aspects and a management part which considers the financial and legal aspects. The first part of the technical portion introduces the reason for the proposed work and clarifies why its solution should be of importance to the potential client. The introduction is followed by a brief statement of the objectives, that is, what the author hopes to be able to achieve by performing the work. This may be followed, if appropriate, by a study of background information, such as the literature surveyed, to indicate that the author is well informed on the subject. Following this, a plan or method of approach is suggested which shows the client that the author has a well-thought-out plan of how he is going to proceed with the work. The method of approach should indicate not only what the author wishes to do, but also what results he expects to obtain from the various portions of his program, what he is going to do if the outcome of the results is as expected, and what if it is not. In conclusion, the technical part of the proposal should include the qualifications of the author or his organization to perform the work.

Although many young engineers may believe that fancy cover designs and big words can sell proposals, it is a fact that the most successful proposals are those that convince the reader of the sincerity and expertise of the writer and his ability to accomplish the objective.

Oral communication

In preparing a speech or oral presentation, first make an outline of the principal ideas that you wish to project. Place them in a logical sequence and prepare your illustrations and similes, but do not attempt to write every word of your speech. Few things are more likely to put an audience to sleep than a speaker who reads his speech. If you tend to be nervous, memorize the first sentence or two, which will get you started, and then use notes only as reminders for the sequence of your talk and to make sure that you have said everything that you wanted to say. Since an audience can best follow simple ideas, it is rarely advisable to present mathematical developments in a speech unless it is to an audience of mathematicians. Nor is it often useful or desirable to delve into the circuitous routes that were used during the development of the idea or the research that is being presented. *The audience is interested in the results and in the usefulness of the results for their own purposes.* All of us are interested primarily in our own life and work, and the better a speaker can convince us that his findings are useful to us, the more successful we believe him to be. Therefore, in preparing a speech, first, find out to whom you will be speaking, and then ask yourself what it is that you can give to the audience that is useful to them. What will they remember after you have stopped speaking?

Your speeches will be successful if you fulfill these four goals of effective public speaking:

1. *Command attention.* (This can be done if you achieve the other three goals.)

There are three things to aim at in public speaking: first to get into your subject, then to get your subject into yourself and lastly to get your subject into your hearers.
—Gregg

2. *Be confident.* (Make a few strong points.)
3. *Know the material.* (Repeat, restate.)
4. *Be enthusiastic.* (An audience will forgive a speaker many things but will not forgive being bored.)

Many things a speaker does or doesn't do may create distraction, causing the audience to miss the main points the speaker is attempting to make. Here are a few hints to remember when making a speech.

Voice: It should be loud enough; vary the pitch, volume, and speed; speak slowly and clearly—but don't drag. Take a tip from the TV commercial—speak loudly the first few sentences. Inexperienced speakers have a tendency to speak too rapidly. They are afraid of pauses, which seem much longer to them than they do to the audience. Speakers are often afraid that if there are moments of silence the audience thinks the speaker has forgotten what he wants to say, or that he doesn't know the material.

Nervous habits: (These are personal mannerisms that are largely subconscious reactions—stage fright.) Don't play with notes, pencil, or pointer. Don't jingle the change in your pockets. Avoid excessive movements of your hands and feet. You may move around a little or change positions and posture, but don't move constantly—or stand statuelike. Many nervous habits express themselves in your language. You've listened to speakers who constantly clear their throats, and you've counted the uh's and aah's, ''you know,'' or some other of the speaker's ''favorite'' words, which he uses to excess.

Poor grammar: Technical talks require, for the most part, the use of formal English grammar and pronunciation. Avoid slang, technical jargon, and long, unfamiliar words when short, simple terms will convey the desired information. A few expressions such as ''we was,'' ''you people setting in the audience,'' and ''ya know'' will cause many in the audience to feel you are less than literate.

Illustration 7-5
Remember, you should not expect that all listeners will be interested in hearing your observations.

Illustration 7-6
If you have a message to tell, be enthusiastic in its presentation.

Eye contact: Look at the audience—look at individuals in the audience—look at their foreheads. Don't look at only one section of the audience or talk to the wall, ceiling, or blackboard.

Gestures: "Elocution," a person's manner of speaking or reading in public, in the early part of this century was a form of highly stylized gestures, head and eye movements, and changes in posture. It was artificially eloquent and theatrical—full of bombast sometimes—but would be considered humorous and old-fashioned today. Be natural! A few gestures are effective, but be careful not to use too many. You may move your body forward for emphasis or sideways for transitions or change of thought. Don't attempt to be "professorial" and hang from or drape yourself over the lectern or podium. Be casual—but not too casual. Remember, all speakers and public performers are somewhat nervous—"excited" may be a more descriptive word, perhaps; just like athletes, they must get "up" for the event. Eventually practice and experience will help you to transmit your excitement to the audience in the proper form—not of nervousness, but as enthusiasm.

The successful speech, like a successful athletic contest, requires practice and rehearsal. In practicing, use a "sparring partner"—a person not afraid to criticize or interrupt and ask questions when something is not clear. Go over a speech with your "sparring partner" again and again, until you are sure that you could present it even if you lost all your notes.

Illustration 7-7
Communication is complex. An idea may be transmitted and acknowledged between earth and moon in a matter of seconds . . .

Illustration 7-8
. . . while a similar transmission and acknowledgment between individuals standing within arm's length may take years.

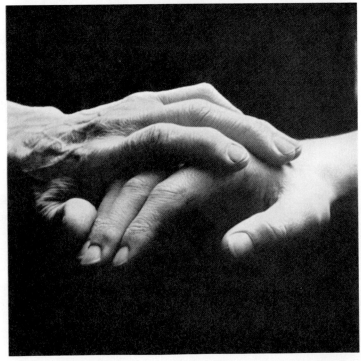

8

The use of computers in engineering and technology

Computers have been available for a number of years but it has been only recently that the individual engineer has begun to use a personal computer extensively. Since many engineering firms are relatively small compared with most industrial and commercial corporations, the initial cost of a large computing system has been difficult to justify. It has only been a result of the recent advances and widespread availability of microcomputers that the smaller engineering firms and the individual have been able to afford their own computer systems. It is this "microcomputer revolution" that has made the computer as much an everyday tool of the engineer as the hand calculator, or even the slide rule of just a few years ago. The computer has been especially useful in analysis type engineering work that was too complex or time consuming for the hand calculator.

One of the most attractive of the computer's capabilities is its ability to perform a large number of mathematical calculations at very high speeds. This capability frees the engineer from having to perform tedious calculations and makes it possible to concentrate on the more important aspects of design that require engineering judgment and experience. With the advent of desk-top computers, many offices that could not previously justify larger units have been able to buy machines that can also assist with bookkeeping, payroll, specification writing, data filing, general typing, and other nonmathematical functions.

Before examining the computer functions that are most useful to an engineering office, we need to understand the components of a computer system. A typical system consists of two groups of components. The physical components that perform computations and transfer information are referred to as *hardware*. The various sets of instructions that are used to direct the function of the hardware are called computer programs or *software*. Hardware falls into three categories: the solid state electronic computing equipment, the documentation equipment, and the equipment that is used to communicate with the computer. This communication process may require several devices that collectively are called the *workstation*.

The workstation

The most familiar device used to send information into the computer is a *keyboard,* Figure 8-1. Computer keyboards are similar to typewriter keyboards except that they usually contain more keys. In addition to the alphabet, numbers, and symbols, computer keyboards usually have keys that are used to send brief or lengthy special instructions to the computer. Such instructions might include directing the computer to perform the necessary calculations in designing a helicopter rotor, or instructing the computer to plot a perspective view of a building.

Another device that may be used to send messages to the computer is a *digitizer.* A digitizer is a smooth-surfaced board ranging in size from several inches (Figure 8-2) to several feet on a side (Figure 8-3). The surface may be either completely blank or more commonly a blank area surrounded by printed symbols and commands (Figure 8-4). The underside has a grid of hundreds of electronically sensitive points.

A digitizer is more graphically oriented than a keyboard. A signal is sent to the computer when a pencil-like stylus or a puck (sometimes referred to as a cursor) is brought close to the electronically sensitive surface of the digitizer (Figures 8-5 and 8-6). The signal may be a specific command to the computer or it may be a record of the position of the puck on the board. Commands to record graphic symbols or to perform certain drafting functions may be sent by positioning the puck over the appropriate location on the surface of the digitizer. In the line-drawing mode the computer records the movement of the puck over the blank area as lines matching the configuration of the puck's path. By mixing linework and symbol signals, information necessary to produce a drawing may be entered into the computer memory (Figure 8-7). This process not only simplifies the generation of technical drawings, but enables the computer very easily to handle any type of two-dimensional data. One of the more popular procedures used in engineering analysis today is based on a numerical technique that divides very large complex systems into many smaller "pieces" that individually are easier to analyze. Each piece can then be evaluated

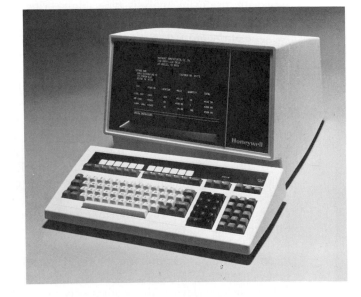

Figure 8-1 A computer keyboard is similar to a typewriter keyboard.

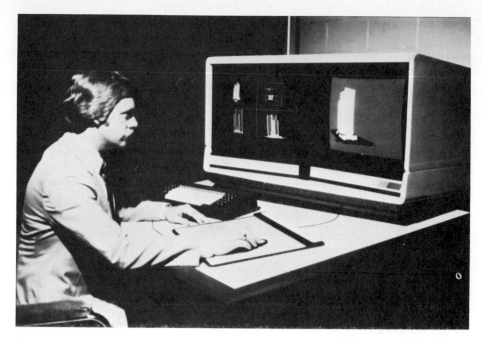

Figure 8-2   A workstation with a small digitizer.

and the results superimposed to predict the behaviour of the entire system. The use of the digitizing tablet can greatly simplify the process of entering the two-dimensional geometry of each one of these pieces into the computer. The layout of the entire system can be placed on top of the digitizing tablet and the coordinate geometry entered by simply placing the puck over the appropriate location on the drawing.

Before discussing a third input device, it is useful at this point to describe how the information going into the computer may be reviewed visually. It is important to see what information is going into the computer to ensure accuracy and to have a base

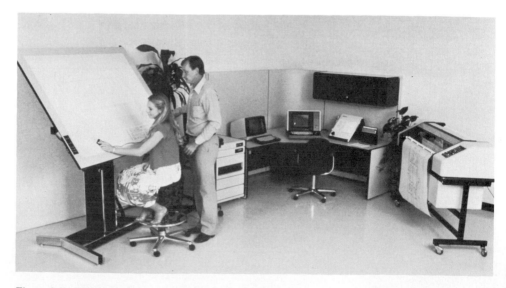

Figure 8-3   This large digitizer resembles a drawing board.

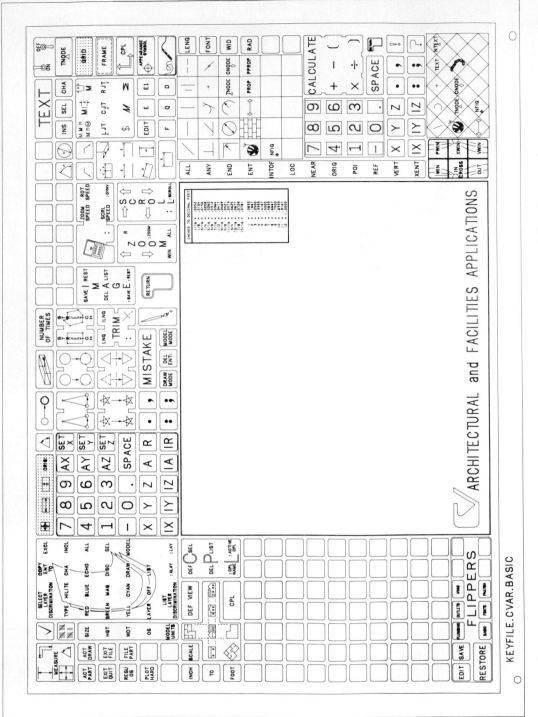

Figure 8-4 The surface of a digitizer board commonly has a blank area surrounded by printed symbols and commands.

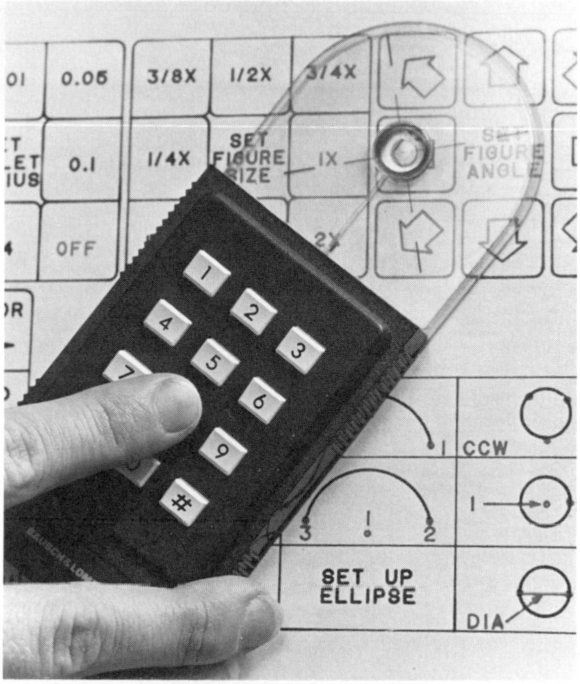

Figure 8-5 A puck may be used to send a command to the computer by positioning it over a symbol on the digitizer.

Figure 8-6 The puck may also be used to record specific locations on the digitizer board.

Figure 8-7 By tracing lines and using symbol signals, a drawing may be entered into the computer's memory.

on which to add more information. A video display *monitor* is used with the keyboard, digitizer, and other input devices to provide a "picture" of the written, numerical, and graphic information being exchanged between the user and the computer. The monitor is a *cathode ray tube* (CRT) similar to a television screen (Figure 8-8). The resolution of the screen varies widely between computer systems. An ordinary TV set may be used as a monitor with many small personal or home computers but the resolution is very limited. The monitors used in large specialized computer graphics systems have much higher resolution. Other types of visual displays, including *liquid-crystal displays* (LCD) like those found in hand calculators and watches, are being developed.

Linework, symbols, letters, and numbers appear on the monitor screen as they are entered through an input device. After the information is seen to be correct, the computer may then be instructed to perform calculations or other functions using the data. The user can then choose either to review the results of the computation on the computer screen or to give the computer additional instructions to perform further calculations. In the larger computer-aided design systems, two monitors are commonly used—one for primarily written and numerical information, and the other for graphical information. In current practice the graphic monitor is frequently a full-color display.

Figure 8-8 The monitor provides a visual record of the written, numerical, and graphic input.

An electronic *light pen* may be used to add or alter information directly on the monitor screen (Figure 8-9). This process is similar to the use of a puck on a digitizer board. The monitor screen may have a group of symbols that, when touched with the pen, will send instructional signals to the computer. The screen also has a blank area on which pen movement will be recorded as lines in the computer.

The workstation is usually located in an ordinary office environment, but the treatment of lighting may need modification. Just as you may see reflections of windows or lamps on your TV screen at home, the images on a computer monitor may be obscured by similar reflections. Outdoor light should not be allowed to fall on the monitor screen. Artificial light must also be controlled in the area where a monitor is located so as to prevent reflections on the viewing screen. Temperature, relative humidity, and sound control need not be different from those normally found in a modern office.

Figure 8-9 A light pen may be used directly on the monitor screen to create and modify drawings and to enter commands.

The computing equipment

The heart of the computer system is the computing unit, whose principal function is to perform many thousands of calculations in a very short period of time. Its essential components are an arithmetic-processing unit, a memory, and a control unit to manage the interactions of these devices and the input and output equipment. In very small computers all of these items may be packaged together along with an input keyboard and possibly even some form of output device. These computers are known as *microcomputers* (Figure 8-10). They are also sometimes referred to as personal or home computers.

Very large computers are known as central or *mainframe* computers (Figure 8-11). Due to their size and special environmental requirements, mainframe computers are usually housed in a special room accompanied by large auxiliary memory units and large high-speed printers. Traditionally, many engineering firms have purchased computer time on a "time-sharing" basis in order to use these larger computer systems.

Between these extremes there are medium-size units called *minicomputers* (Figure 8-12). These units are often installed in the same room as the workstation. Due to the great advances in solid-state electronics in recent years, the calculating power of many microcomputers is now very close to that of the minicomputers of just a few

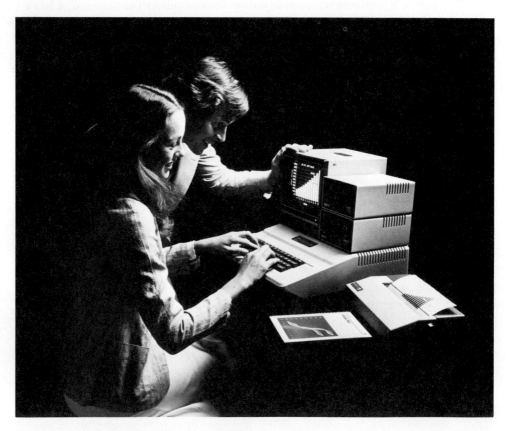

Figure 8-10 Very small computers . . . known as microcomputers or personal computers . . . have their essential components combined in one package.

Figure 8-11 Very large computers . . . known as central or mainframe computers . . . are usually housed in a special room.

Figure 8-12 Medium-size computers are called minicomputers.

years ago. Furthermore, many of today's minicomputers have surpassed the capabilities of many of the older mainframe computers as well.

If very extensive graphic applications are desired, engineering offices are most likely to use a minicomputer or a large microcomputer system. At present, the minicomputer is the most widely used system for graphics, but as better graphics software (instructions that tell the computer what to do) is developed for the larger microcomputer systems, this will undoubtedly change. If more extensive computing capability is also required, a network of computer workstations can be used. Microcomputer systems that can function individually as stand-alone computer systems and as workstations for a single minicomputer system provide a tremendous amount of computing flexibility. Routine calculations and graphic manipulations (especially in regard to input and output) are processed by each microcomputer. More lengthy calculations and expanded memory requirements are shifted to the minicomputer. If the engineer does not require extensive graphics capability, microcomputers alone are most often used. Many of the personal computers available today are capable of meeting most of the daily computing requirements of an average engineering-consulting firm.

The *memory* unit is a very important part of the computing equipment. It is used to store the data to be processed, the final and intermediate results of any calculations, and the software instructions that dictate how the calculations are to be made and displayed or printed. The memory unit also serves as a library in which drawings, specifications, letters, and accounting information can be stored. Commands entered at the workstation will cause the computer to perform a series of tasks previously entered into memory. The tasks might include analyzing machine parts for internal stresses, generating mechanical drawings, writing specifications, and a multitude of other mathematical and nonmathematical procedures.

In order to create a drawing on a computer, two basic types of data must be stored in memory. First is the drafting program of instructions that tell the computer how to make the drawing. Although large, the magnitude of the data *is not* related to the number of drawings in the system. The second type of data is that developed by the program to define the subject material on each drawing. The size of the data in this case *is* related to the number of drawings in the system as well as to the size and complexity of each drawing.

About 25,000 words are needed to store the instructions necessary to create (draw) an average architectural or engineering drawing. A set of thirty drawing sheets would require the computer to store at least 750,000 words—too many for the built-in memory of a typical microcomputer. Therefore an auxiliary storage system is needed. Minicomputers and mainframes also use auxiliary memory units to increase their capacity as needed. *Magnetic tape, hard discs,* and *floppy discs* are common auxiliary memory devices. Punched cards and paper tape have also been used with some of the older systems.

To store completed projects and other information not regularly used, "off-line" storage devices must be provided. This is to allow new projects to be entered into the system because on-line memory has a finite capacity. Floppy discs are used with microcomputers and magnetic tape is used with larger computers for off-line storage. Auxiliary memory storage systems give the computer a much greater capability than if it could only store programs and data within its central processing unit.

The computing equipment's output (results of computations and other performed tasks) is displayed on a monitor at the workstation throughout the exchange between the user and the computer. To obtain a printed copy of the output, documentation equipment is required.

The documentation equipment

The display of the conclusion of the computing process (charts, graphs, perspectives, floor plans, lists, structural designs, and so on) has limited use if it is merely shown as a picture on a CRT screen. The output of most computer work must be printed on paper to be a useful tool in any engineering office. Computer systems therefore also include machines that can print and plot written and graphic results. Small desk-top computer systems usually have their printing equipment located adjacent to the workstation. Larger computers and their associated high-speed printing equipment often make enough noise to be bothersome. They are therefore best located in a room separate from the general work area. It should have sound-deadening properties.

Computer printing equipment is usually referred to as *printers* or *plotters,* depending on primary use. *Printers* are designed primarily for alphanumeric (letter and number) output similar to that of a standard typewriter. Some types of printers can produce graphical output as well. In this case, rather than having individual alphabet and number elements (like a typewriter), letters and numbers are formed by a series of tiny dots. The dots can also be used to form line and graphic symbols. Other types of printers are designed primarily to print graphical output. They can produce alphanumeric output as well but are not designed for high-speed printing of large quantities of small, standard print. Lines drawn by plotters are usually continuous rather than being formed by a series of dots.

Electromechanical *pen plotters* employ several types of pens and two types of plotting surfaces. Pencils, ballpoint pens, and ink pens may be used as marking devices. The quality of the drawing increases from the first to the last of these. The mechanism that holds the paper may be flat (*flat-bed plotters*), as in Figure 8-13, or

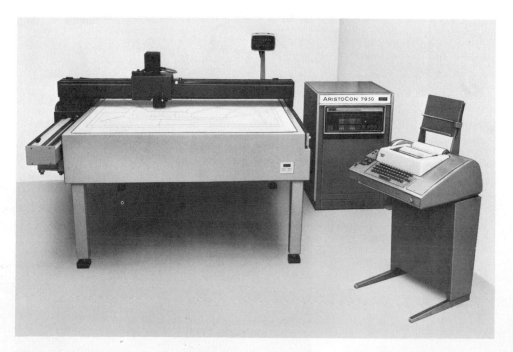

Figure 8-13 A flat-bed plotter holds the paper on a flat surface . . . it may also have digitizing capabilities.

cylindrical (*drum plotters*), as in Figure 8-14. Flat-bed plotters are available in a variety of sizes corresponding to standard drawing sizes. The pen travels left and right on a horizontal bar that moves up and down on a flat drawing surface. Drum plotters print on paper in contact with a cylindrical drum. If the paper is fed from a roll, the drawing can be quite long. The width of the drum may vary from the width of standard typing paper to that of large drawing sheets. The drum holding the paper rotates in both directions so that a pen mounted on a fixed bridge can draw lines in the direction of rotation of the drum. The pen also moves back and forth on the bridge to draw lines parallel to the axis of the drum. The combination of these two movements enables lines to be drawn at any angle or curvature.

Computer-drawn linework is not always drawn in the sequence that a drafter would follow to complete a sheet. Since the computer memory contains all the data necessary to produce a drawing, the lines, notes, symbols, and numbers may be applied to the drawing sheet in a more efficient sequence. A drum plotter, for example, may plot along one end of a sheet and as the drum rotates back and forth it will gradually roll the sheet past the pen bridge so that the drawing is completed from one end to the other. In contrast, the drafter usually applies object lines first and then finishes with notes and dimensions. It is no advantage for the computer to print notes forward (the first word first and the last word last). The computer output program may be designed to print all data based on the position of the pen at a particular moment. This may result in notes, dimensions, and linework being plotted both from left to right and right to left.

A variety of plotter linework styles may be readily selected (Figure 8-15). Colored

Figure 8-14 A drum plotter holds the paper on a rotating drum.

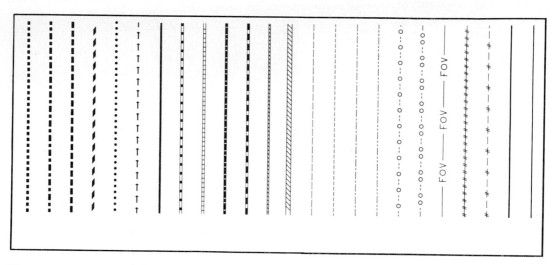

Figure 8-15 A variety of plotter linework styles may be selected.

lines may also be drawn by plotters that hold several pen points. A variety of lettering styles is also available (Figure 8-16). Many symbols can be printed without a drafter forming them individually on a digitizer. The symbol picture on the digitizer need merely be touched with the cursor for the image to be printed on the drawing sheet (Figure 8-17). The size of the symbol library and the ease in which new sym-

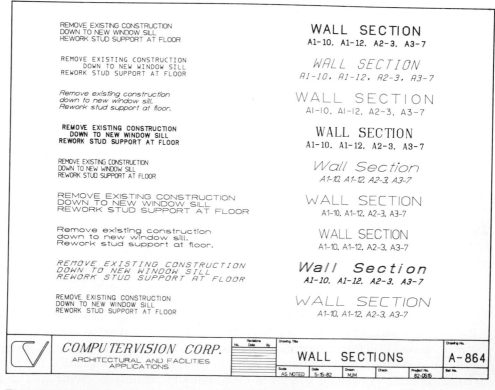

Figure 8-16 A variety of lettering styles may also be selected.

Figure 8-17 Symbols may be printed directly on a drawing by touching the desired symbol picture on the digitizer.

bols may be drawn and stored for future use varies greatly among the many graphic software and computer systems.

A faster printing device that does not use pens is the *electrostatic plotter*. The quality of printing with this type of plotter has generally been poor but is now improving. This machine utilizes hundreds of fixed electronically sensitive points to place dots on paper. As the paper moves past the long bank of dot-printing styli, thousands of dots are printed to form lines, letters, numbers, and symbols.

Computers and printing equipment work best in an environment of controlled temperature and humidity. Temperatures in the 60 to 75°F range and relative humidity from 35 to 70 percent will limit expansion and contraction of component parts and drawing paper. Maintaining these conditions contributes to smooth functioning and accuracy in drawing. Rapid temperature changes should also be avoided. Since the equipment generates heat, this problem must be considered when designing the air conditioning for the space.

Equipment selection

The various components of a computer system may be purchased separately or as a package. When purchasing separate items, it is very important to make a careful analysis of their compatibility. This is especially true of a system having graphic capabilities. Some manufacturers supply graphic computer systems only as a complete package of hardware and software. These "turnkey" systems are closely integrated to perform a specific design function and updated as new developments arise.

Another point to keep in mind in equipment selection is that there are major differences between systems in their ability to "multitask." Some systems can only perform one function at a time, such as plotting a drawing, creating a drawing on a monitor, or digitizing a drawing. Others can do more than a single task at one time. This ability has a major impact on productivity and scheduling in an engineering office.

Other aspects that must be considered are:

Availability of software

If a system is intended to be used with a specific piece of software, always check to be sure that it is available for that specific system. Too many computer systems are idle because of the lack of the appropriate software to do the job. If a system is being selected with software development in mind, items such as software compatibility between different systems and the availability of the necessary programming languages (BASIC, FORTRAN, PASCAL, etc.) should be considered.

Compatibility with other systems

Many times a computer system is required to perform operations on data that are generated by another computer system. If the format of the external storage media (floppy disks, hard disks, tape) for each system is different, then compatibility is not easily accomplished. This problem is particulary acute when you are working with microcomputers.

Support

Be certain the supplier can provide the necessary support, both to the computer system and any selected software. This is again a problem that surfaces primarily with microcomputers. If the supplier is not a licensed dealer for a particular system (i.e., a mail order house), it is likely that adequate support will not be provided.

Computer software

Before any computer system can be put to practical application, the necessary sets of instructions or *programs* must be written or purchased to direct the computing and associated equipment. The computer manufacturer usually supplies only the most elementary programs with the purchase of the computer. Other specialty programs must be carefully selected and purchased to suit your specific needs.

As mentioned previously, the availability of software varies from one computer system to the next. It is therefore very important when selecting a computer system to look carefully at both the hardware and software capabilities of each manufacturer's products. Some special purpose computer programs have also been written by educational institutions and by companies that are not directly associated with any one hardware manufacturer. Programs of this type are usually designed to run successfully on the most popular models of computers from several major manufacturers with only minor modifications.

No matter how well a computer program is written, a certain amount of time must be allocated for training operators before productive use can be expected. Software (as well as hardware) that is easier to learn to use is said to be "user friendly." Training courses are often provided by manufacturers or software suppliers at their headquarters or at your office when a large purchase is made. Additional information and modifications to widely used programs are available by joining "users' groups" that share practical application techniques.

In the following sections we will briefly look at the general application areas of computer systems in engineering design offices. These areas are design calculations, word processing, computer-aided drafting, and computer-aided design. Many programs are commercially available in each of these areas.

Design calculations

Until recently, the computer has had limited use in the area of routine design calculations. This has been primarily due to the way computers have been used by engineers in the past. Usually, large mainframe or minicomputer systems do not lend themselves to the daily process of generating design calculations. The restrictions imposed by cost, access (often the larger computer systems must be accessed via a specialized phone line in an area away from the designer's desk), and the lack of good software are several good reasons why the computer has not been routinely used for this type of work. Also generating design calculations is a "real time" process, and the inconveniences imposed by the inflexibility of the larger computer systems have severely limited the computer's use. This problem has resulted in the computer being used only in situations where special attention must be given to specific aspects of the design. Once the computer has solved these specialized problems, the design engineer returns to the traditional pencil and paper process of generating design calculations. Up to now this procedure has been fairly common; but, with the increasing development of the microcomputer, practices will undoubtedly change. Electronic "spread sheet" programs developed specifically for the microcomputer enable the design engineer to lay out the design calculations directly on the computer's screen, with the computer performing all the numerical calculations. This procedure will allow the engineer to focus on more important apsects of the design and be freed from the sometimes tedious and error-prone process of hand calculation. The design process is essentially an iterative one in which the engineer must consider many different options before arriving at a final design. Ideally, these alternatives are optimized in such a way that the best design is achieved in terms of cost and reliability. Through the use of a spread sheet program like the one described above, the engineer can achieve optimization in a much faster, more convenient, and efficient manner.

Word processing

The previously tedious and inefficient task of cutting up old specifications and pasting them together with newly written segments is now made relatively simple by the computer. Computer programs are available with complete specifications on them. The specification writer may delete or add segments to the computer program without handling paperwork. This procedure not only makes the preparation of specifications faster but also makes the document more accurate. There is less chance of omitting critical data or including extraneous or dated information. Once the specifications are complete, the computer prints them, thus saving hours of typing.

Bookkeeping, a function necessary in all offices, can also be computerized. Billing and payroll data can be stored on the computer. Such information can be called up on the monitor, updated, and returned to the computer memory by pushing a few buttons on the keyboard at the workstation. Meeting notes, correspondence, and other general information formerly kept in a file cabinet can be stored on the computer instead. Also, the routine task of typing letters can be done on the computer. Correction of typographical errors and general editing can be done on the monitor before the computer proceeds to type the text on paper. Changing typed material after the text is printed simply means that the original text is retrieved from the computer's memory, displayed on the monitor, edited electronically, and then a fresh copy is printed. The obsolete copy itself is not corrected. It is usually discarded.

Computer-aided drafting

Like the design process, the preparation of technical drawings is repetitive. Drawings may be generated at each stage of the design, from the initial design (where rough design assumptions are made and then improved upon) to the final design (where the precise features of the design must be reflected in the final drawings). It is here that computer graphics capability makes its greatest impact. Complete drawings can

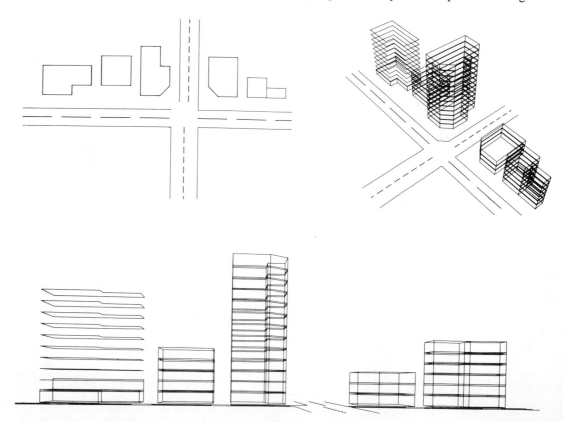

Figure 8-18 Plan and elevation data entered in a computer can quickly produce perspective views from various directions.

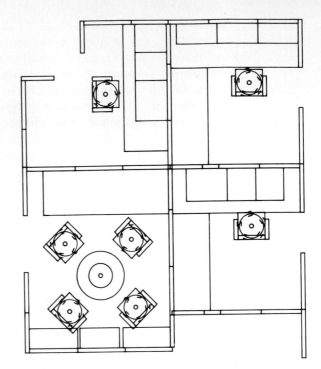

Figure 8-19 Pictorial views may also be drawn in isometrics by the computer.

be stored in the computer and updated in the same way that a secretary would update a document using a word processor. In addition, technical drawings from different areas of the design can be overlaid (such as a structural and a mechanical drawing) directly on the computer screen and any conflicts between the two can be resolved graphically. Computer graphics capability not only adds a tremendous amount of flexibility, but once typical design details are entered into the computer, a tremendous amount of time can be saved by simply reusing the same detail from one design to the next. It should be noted that the preparation of drawings in any technical field follows essentially the same process. The process of constructing perspective drawings in the architectural engineering field is a good example. The distance of the

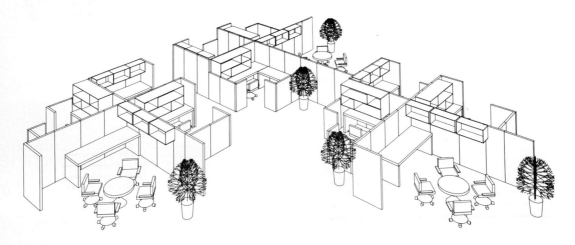

Figure 8-20 A computer-drawn interior perspective.

viewer from the building, the height of the viewer, the angle of the view, and even the sides of the building to be viewed all affect the form of the perspective drawing. Although an experienced drafter can readily make such decisions, considerable work is invested before one can tell from the drawing in progress just what will be seen in the final rendering. Often, when working by hand, only one or two perspectives may be drawn because of the labor required. This limits the designer's and the client's ability to judge what the constructed building will look like from all directions. The graphic capability of computers has drastically changed this limitation, as shown in Figures 8-18 to 8-20.

Computer-aided design

Thus far the computer uses described have been essentially those replacing manual activities. The design capability of a computer may seem to emulate the thinking process of a designer. However, it operates the same for design decisions as it does for other activities. In designing, the engineer can type into the computer the criteria for a specific design. As mentioned previously, the computer can then determine the optimum utilization of the criteria and report on the monitor the preferred solution. The process is one of trial and error, wherein the computer proposes to itself different options and then evaluates them mathematically on the basis of the priorities set by the designer.

The great speed of the machine in making such evaluations makes it seem as if the machine is designing, but it is merely reporting an optimum mathematical relationship. This ability can save the designer time in developing complex designs, but certainly it is not the complete design process. The computer tool is valuable in that it can save the designer time in certain of the routine sequences of operations and decisions that are typical of the design process.

When using the design capabilities of the computer, it is always necessary to recognize the importance of selecting the most appropriate computer software that will perform the analysis required. It is also very important to understand the assumptions on which the selected software is based. Being able to evaluate, interpret, modify, and write computer programs will be important design functions for the engineers and technologists of the future.

Conclusions

As computer systems continue to become more and more attractive in terms of both cost and capability, the daily process of producing work in engineering offices will surely change. Professionals must continue to understand the principles and concepts of engineering design, but will ultimately perform very different tasks in effecting these designs. More pushing of pens on screens and less pushing of pencils on paper is no doubt in the future of most technical professions. Before the computer can be more than just another tool, however, more work will have to be done to reduce engineering judgment and experience to a mathematical basis. Since these qualities are difficult to describe verbally, not to mention numerically, they will remain the prerogative of the human mind.

Problems

8-1. Explain the function of a digitizer.

8-2. Visit a local engineering office and list the types of computer equipment used. What is the function of each? Is the plotter flat-bed or drum type?

8-3. What is a light pen?

8-4. For personal use would you purchase a microcomputer or a minicomputer? Why?

8-5. Describe the operation of the memory unit of a digital computer.

8-6. What is a floppy disk?

8-7. Visit a local engineering office that uses computer-aided drafting equipment. What software is used?

8-8. What are the advantages and disadvantages of word processing?

8-9. What are the advantages and disadvantages of computer-aided drafting.

8-10. How has the computer improved the design proficiency *potential* of the average engineer?

9

The modeling of engineering systems— mechanical, electrical, fluid, and thermal

The basic work of most engineers is concerned with the design and analysis of physical systems. This process is aided by conceiving of a total system as an arrangement of component parts and then representing these components as idealized elements which possess important characteristics of the *real* components and whose behavior can be described. It further involves the expression of the behavior of the different parts of the system and, finally, of the system as a whole in mathematical terms.[1] These two steps are called modeling and they are among the most important tasks performed by today's engineer.

In the modeling process the engineer's first task is to devise a conceptual model that schematically represents the real system being studied. The next step is to express this conceptual model mathematically. This expression of physical components by mathematical forms is referred to as developing a "mathematical" model. Once an engineer has chosen a mathematical model of a real system, it is possible to predict how the system would behave in various conditions and under different constraints without actually building sample systems and testing them experimentally. They can be a very slow and expensive process. In other words, one can use a mathematical model to analyze and revise a particular system design.

> When we mean to build, we first survey the plot, then draw the model; And
> when we see the figure of the house, then must we rate the cost of the erection.
> —King Henry IV, Part II, Act I, Sc. 3, Line 41

[1] A more extensive treatment of the material presented in this chapter can be found in J. L. Shearer, A. T. Murphy, and H. H. Richardson, *Introduction to System Dynamics* (Reading, Mass.: Addison-Wesley Publishing Company, Inc., 1967).

The construction of a mathematical model of a real system is usually based on the relations between the *inputs* and *outputs* of certain idealized basic elements which make up the chosen conceptual model. As an example, in mechanical systems, a spring is a commonly used element. The engineer expresses mathematically the relation between a force applied to the spring (input) and the resulting displacement or stretching of the spring (output). Other elements commonly used are electrical resistances, mechanical dampers, fluid capacitors, etc. Finally, the engineer must combine all the basic elements so that mathematical equations can be obtained which relate the system inputs to the system outputs.

The engineer must understand the concept of system *inputs* and *outputs*. Figure 9-1b is a schematic representation of system inputs and outputs. If an aircraft is the system that we are attempting to model, the *inputs* to the system would be such factors as wind gusts, throttle position, and the position of the aircraft control surfaces, such as the elevator, aileron, and rudder. *Outputs* from the system that the engineer would like to be able to *predict* would include such quantities as the aircraft's altitude, speed, and acceleration; that is, the *inputs* act on the system to produce the *outputs*.

Figure 9-2 provides an example of the modeling process for each of the four basic types of systems: electrical, mechanical, fluid, and thermal. In each case the actual physical system is first conceptualized as an idealized physical model.[2] Then this idealized model is represented by a mathematical model. In other words, the action performed by the system, or the relation between its input and output, is finally described in terms of mathematical equations.

If we were modeling the vertical dynamics of an automobile as it travels down a road, an input to the actual system would be the roughness of the road. An output might be the acceleration felt by the passenger in the car. The conceptual model of

[2] The symbols and schematic that are used to represent the various elements will be explained in a later section.

Figure 9-1

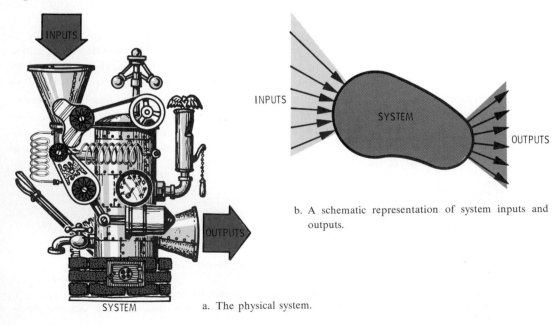

a. The physical system.

b. A schematic representation of system inputs and outputs.

Figure 9-2 Examples of the modeling of physical systems.

The Physical System An Idealized Conceptual Model A Mathematical Model

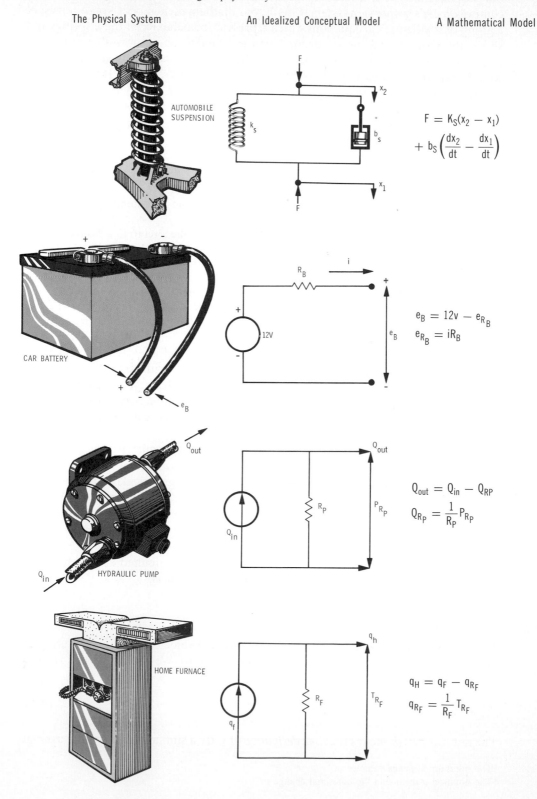

AUTOMOBILE SUSPENSION

k_s b_s

$$F = K_S(x_2 - x_1)$$
$$+ b_S\left(\frac{dx_2}{dt} - \frac{dx_1}{dt}\right)$$

CAR BATTERY

R_B i

$12V$

e_B

$$e_B = 12v - e_{R_B}$$
$$e_{R_B} = iR_B$$

HYDRAULIC PUMP

Q_{out} R_P P_{R_P} Q_{in}

$$Q_{out} = Q_{in} - Q_{RP}$$
$$Q_{RP} = \frac{1}{R_P}P_{R_P}$$

HOME FURNACE

q_h R_F T_{R_F} q_f

$$q_H = q_F - q_{R_F}$$
$$q_{R_F} = \frac{1}{R_F}T_{R_F}$$

our suspension system would be the series of springs and shock absorbers which comprise a car's suspension system. Finally, our mathematical model would consist of a group of mathematical equations which represent the behavior of these springs and shock absorbers. By means of our mathematical model we are now able to test the suspension system that we have modeled by mathematically determining how it would respond to a variety of irregularities in the road surface. Hence, we can determine whether our original system will behave properly. If its performance varies from our expectations, we can easily make changes in our conceptual model, which will hopefully improve its behavior. Finally, we can actually build a new physical system which resembles our revised model.

Problems

9-1. What is the primary input and the primary output of an automobile power steering system?

9-2. The purpose of one type of aircraft autopilot is to return the aircraft to a desired altitude if it is forced up or down by wind gusts. The autopilot works by sensing altitude changes and moving the elevator surface automatically to return the aircraft to the desired altitude. What is the input and output of the autopilot?

9-3. Your home heating system consists of a thermostat that is sensitive to the difference between the room's temperature and some preselected desired temperature and a furnace that produces heat energy. Define the input and output of this system.

In this chapter the systems that we will learn to model are "dynamic" systems. This implies that their inputs and outputs are changing with time. Generally the equations by which we describe the behavior of physical systems will contain *variables* and their *time derivatives*. These are called *differential equations*. For simple systems these differential equations can sometimes be solved with pencil and paper, but often their solution requires the use of computers.

A concept that is inherent in the study of system dynamics is that of "rate of change." Let us consider a system which involves variables and their time derivatives. An automobile traveling over a bumpy road has a changing vertical position, velocity, and acceleration. In mechanical systems we define the *velocity* of a point as the time rate of change of the point's *position*. Similarly, *acceleration* is defined as the time rate of change of the point's velocity. In graphical form, Figure 9-3 illustrates how a point's position (s), velocity (v), and acceleration (a) will vary with *time*. The average velocity during a particular time interval ($\Delta t = t_2 - t_1$) is defined as the change in position ($\Delta s = s_2 - s_1$) divided by the time interval; that is,

$$v_{av} = \frac{s_2 - s_1}{t_2 - t_1} = \frac{\Delta s}{\Delta t} \qquad (9\text{-}1)[3]$$

The *instantaneous* velocity of the point is the average velocity of the point during an *infinitesimal* time interval, that is, as Δt approaches zero. We will define this velocity as

$$v = \lim_{\Delta t \to 0} \frac{\Delta s}{\Delta t} = \frac{ds}{dt} \qquad (9\text{-}2)[4]$$

The notation ds/dt means the *time derivative* of s. By a similar argument the average

---

[3] The notation Δ means "change in."

[4] The notation d means an "infinitesimal change in."

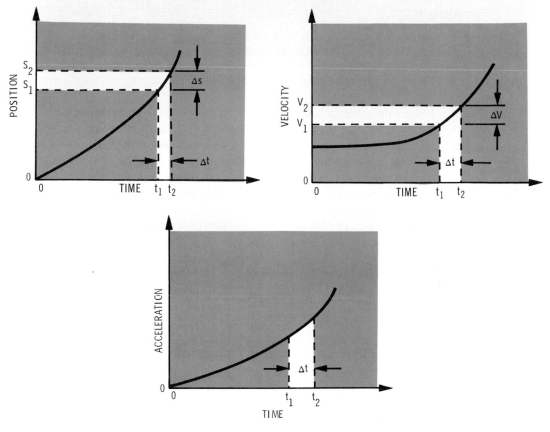

Figure 9-3 Concept of time rate of change.

acceleration is defined as the change in velocity divided by the change in time,

$$a_{av} = \frac{v_2 - v_1}{t_2 - t_1} = \frac{\Delta v}{\Delta t} \qquad (9\text{-}3)$$

and the instantaneous acceleration is the *time derivative* of the velocity,

$$a = \frac{dv}{dt} \qquad (9\text{-}4)$$

Consider a driver traveling from Paris to LeMans. At a given instant his speedometer reads 72 km/hr, his watch reads 3:30, and his odometer reads 64,360 km. Ten minutes later he notes that his odometer reads 64,375 km, his speedometer reads 88 km/hr, and his watch reads 3:40. Using equations 9-1 and 9-3, we can compute his average velocity and acceleration. For example, his average velocity is

$$V_{av} = \frac{(64{,}375 - 64{,}360)\ \text{km}}{10\ \text{min}} \times \frac{60\ \text{min}}{\text{hr}} = \textbf{90 km/hr}$$

and his average acceleration is

$$a_{av} = \frac{(88 - 72)\ \text{km/hr}}{10\ \text{min}} \times \frac{60\ \text{min}}{\text{hr}} = \textbf{96 km/hr}^2$$

Problems

9-4. In Figure 12-3, $s_1 = 300$ km, $s_2 = 350$ km, $v_1 = 70$ km/hr, $v_2 = 90$ km/hr, $t_1 = 3$ hr, and $t_2 = 3$ hr 15 min. Calculate the average velocity and average acceleration.

9-5. A rock falling freely from rest off a mountain is observed to have a velocity of 2.45 m/s after 0.25 sec. Find the average acceleration of the rock during this time interval.

By means of mathematical expressions known as differential equations we can write equations which will describe the behavior of models of dynamic physical systems. We are now familiar with the basic elements used in the modeling process: the conceptualization of the physical components of the system, and the development of a mathematical model by which they can be described. We are now ready to consider a more extensive example of how the modeling process can be used in solving a real engineering problem.

By 1990, it has been predicted that high-speed ground transportation vehicles may exist that travel as fast as 500 kilometers per hour (Illustration 9-1). At these speeds even small irregularities in the guideway can produce uncomfortable accelerations in the passenger cabin. Therefore, the vehicle's suspension system must be carefully designed to (1) support the vehicle's weight, and (2) isolate the passenger cabin from excessive vibrations.

At the start of such a major design project, it is not feasible to construct scale models of a number of possible designs and test them experimentally. However, it is possible to *model* a system conceptually and mathematically, as illustrated in Figure 9-2. The first step in this process is to formulate a conceptual model of an actual system. This can be accomplished by assembling various masses, springs, and dampers which, together, will resemble the behavior of the actual system. A simplified conceptual model of a design that could meet the two stipulated design requirements is shown in Figure 9-4.

The second step is to formulate a mathematical model that will represent the characteristics of each component in the conceptual model. In this case the input to the mathematical model is the ground irregularity, the output is the passenger compartment acceleration, and the mathematical model itself is the particular set of mathematical expressions (differential equations) that will include the suspension and mass components. Once this mathematical model has been formulated the designer

Illustration 9-1
High speed ground transportation systems can be modeled conceptually using arrangements of masses, springs, and dampers.

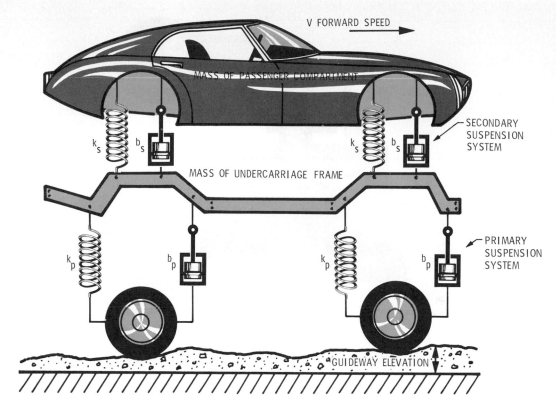

Figure 9-4 Conceptual model of a vertical suspension design.

can vary the characteristics of each system component until he is satisfied with the *performance* of the system as a whole.

In some cases our engineer may decide not to use springs and shock absorbers at all. For high-speed ground systems it has been proposed that a noncontacting suspension such as an "air cushion" suspension be used. Figure 9-5 is a schematic sketch of the fundamental components of such a system. In order to evaluate the feasibility of such a *fluid* suspension system, we will need to formulate a new conceptual model. Notice that in this new system the conceptual model will involve fluid component relationships such as the *volume* flow through the flow control valve and the *pressure drop* across it. It will be important to determine the relationship between the system inputs, such as the road irregularities, and the system outputs, such as the vertical acceleration of the passenger compartment.

Clearly the fluid suspension in this design is performing the same *function* as would a mechanical spring and dashpot[5] suspension. We will find that this similarity applies not only to the function but also to the mathematical modeling. In fact, each of the four basic types of physical systems (mechanical, electrical, fluid, and thermal) can be used to perform similar functions and operate in analogous ways. More than at any time in the past, the modern engineer must have a broad understanding of all physical systems and not limit his or her knowledge to one or two specialized areas. The successful practicing engineer—the group leader, the section chief, the chief engineer—must understand, for example, how energy is transferred between mechanical, electrical, thermal, and fluid systems and must be able to design interfaces

---

[5] A dashpot is a mechanical damper, like the shock absorber on a conventional automobile.

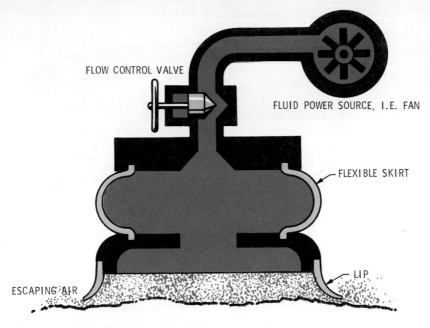

FLOW CONTROL VALVE

FLUID POWER SOURCE, I.E. FAN

FLEXIBLE SKIRT

LIP

ESCAPING AIR

Figure 9-5 Schematic sketch of air cushion suspension.

between these different media. Also, an understanding of the basic and underlying analogies that exist will enable the specialist, educated primarily in one field, to apply that knowledge to other fields.

When an engineer wants to transfer energy from one type of system to another, many familiar devices are available. An electric motor transforms electric energy into mechanical energy; an electric generator transforms mechanical energy into electrical energy; a hydraulic actuator transforms fluid energy into mechanical energy; a hydraulic pump converts mechanical energy into fluid energy; and a steam engine converts heat energy into mechanical energy. All these devices are called *transducers* because they transfer energy from one media to another, and they are used in many engineering applications.

To perform many jobs, the engineer must be familiar with all four basic types of physical systems: mechanical, electrical, fluid, and thermal. However, because there are basic conceptual similarities between all four, it is not necessary to learn a separate set of rules for dealing with each. By understanding their underlying similarities, specialized knowledge for one type of system can be applied to any of the others.

Some of the analogies between real physical systems are familiar to us all and are quite simple to understand. Each of the four basic types of systems can be conceived of as possessing "through" and "across" characteristics. We can think of current flowing *through* a wire, liquid flowing *through* a tube, force being transmitted *through* a spring, and heat flow (thermal energy) being transmitted *through* a conductor. Electric current, fluid flow, mechanical force, and thermal heat flow are both physical and conceptual analogies. We will call them "through" variables because they are measured in terms of a quantity transmitted *through* a respective element.

Each of these four basic types of systems also possesses "across" characteristics. Electric voltage, fluid pressure, the velocity of mechanical elements, and temperature are analogous in that they all are measured between or *across* two terminals, one at the "entrance to" and one at the "exit from" their respective systems. The "across" variables—voltage, pressure, velocity, and temperature—have a different value at

Figure 9-6 Conceptual model of a mechanical system.

each terminal. It is this difference that we measure. When we speak of a 12-volt battery, we mean a battery in which we can measure a 12-volt difference between or *across* the two battery terminals.

When we talk about the value of an "across" variable, we need to define a reference value. Generally this reference value is arbitrary but constant; that is, voltage is measured with respect to "ground," velocity is measured with respect to a fixed (generally zero) velocity, pressure is measured with respect to a zero pressure (absolute) or atmospheric pressure (gauge), and temperature is measured with respect to the zero point on a fixed scale (0° Celsius, 0° Fahrenheit, 0° Kelvin, or 0° Rankine).

To illustrate the difference between "through" and "across" variables, let us look at a hypothetical model of a simple physical system. Figure 9-6 could represent either a mechanical, a fluid, an electrical, or a thermal system. In Figure 9-6 the "through" variable f (current, fluid flow, force, or heat flow) is shown flowing through the element from terminal 2 to terminal 1. The "across" variable v (voltage, pressure, velocity, or temperature) is a measure of the difference between the values of v_2 and v_1 at the two terminals.

If Figure 9-6 represented an electrical system, f would represent electrical current, and $v = v_2 - v_1$ would represent voltage. If Figure 9-6 represented a mechanical system, f would represent force, and $v = v_2 - v_1$ would represent velocity. If Figure 9-6 represented a fluid system, f would represent fluid flow (volume flow rate), and $v = v_2 - v_1$ would represent the pressure difference. If Figure 9-6 represented a thermal system, f would represent heat flow and $v = v_2 - v_1$ would represent the temperature difference between the terminals.

This basic concept of the difference between "through" and "across" variables and their analogous roles in all four types of systems will prove to be a very useful tool for us in understanding physical systems, for it will allow us to apply our knowledge of one type of system to any of the others.

Elements of dynamic systems

Now let us develop the basic building blocks of model dynamic systems. In general these building blocks will be idealizations or approximations of physical realities. However, this lack of precision will be compensated for by increased model simplicity. Therefore, we will model our systems using *ideal elements*, that is, elements that are represented as simplified mathematical relationships. More complex models should be utilized in those cases where the simpler models do not adequately describe the phenomena being investigated.

Power and energy

In general, the ideal elements that we will consider will store, dissipate, or transform *energy*. Energy can be defined as the ability to do *work*, and *power* is defined as the *time rate* of doing work. Equivalently power can be defined as the rate at which energy *flows* into or out of an ideal element. If we define the energy contained in an ideal element as E, then the power is $\mathbf{P} = dE/dt$, where positive ($+$) power represents energy flowing *into* the element.

Choice of dynamic variables

Earlier in this chapter analogies were made between mechanical, electrical, fluid, and thermal systems by defining *through* and *across* variables. If an engineer has a good understanding of any one of the different media, then by understanding the analogies between systems it is possible to transfer knowledge obtained for one specific use to other media. To make this easier, we will describe each ideal element of the various media in terms of its *through* and *across* variables.

Traditionally *through* and *across* variables have been called *power variables*[6] because, in all except thermal systems, the product of an element's through and across variables is the power delivered to the element. If f represents a generalized through variable and v represents a generalized across variable, the power is

$$\mathbf{P} = fv$$

In a mechanical system, power is the product of force and velocity; in an electrical system, power is the product of current and voltage; in a fluid system, power is the product of volume flow rate and pressure. In thermal systems power is simply the heat flow rate. The units of power are FLT^{-1}. It is clear from this expression that the fundamental dimensions of power will not vary from system to system. The SI unit of power is the watt, which is a joule/second or a newton-meter/second.

Mechanical systems

In this section we will analyze mechanical systems in which the motion of the system remains parallel to fixed axes. The ideal elements that will be most useful in analyzing mechanical translational systems are the ideal mass, spring, and damper.

Ideal translational mass

Newton's Second Law of Motion expresses the relationship that exists between a force (f_m) that is applied to a mass and the time rate of change of the mass's *linear momentum*,[7] dp_m/dt, that is,

$$f_m = \frac{dp_m}{dt}$$

where p_m is the linear momentum of the mass. An *ideal* mass is defined as one in which the linear momentum is linearly proportional to its velocity,

$$p_m = mv_m \tag{9-5}$$

where v_m is the inertial *velocity* of the mass, that is, with respect to a "fixed" reference.

---

[6] There are other variables that can be used to describe dynamic systems; however, the analogies between systems are more easily understood by the use of power variables.

[7] The rate of change of momentum of a body is proportional to the resultant force acting on the body and is in the direction of that force.

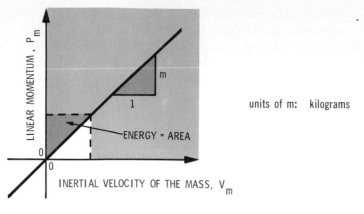

Figure 9-7 Momentum-velocity relationship for an ideal mass.

Substituting this ideal relationship into Newton's second law yields the familiar result,

$$f_m = m\frac{dv_m}{dt} = ma_m \qquad (9\text{-}6)$$

where dv_m/dt is the inertial acceleration of the mass. Figure 9-7 shows this linear relationship for an ideal mass.[8] Note that for ideal elements the energy can be found from either the area above or below the curve, as indicated. For non-ideal elements these areas are not equal and considerable care must be taken to choose the correct one. However, a consideration of non-linear elements is beyond the scope of this chapter.

Figure 9-8 is a conceptual model of a translational mass. On conceptual models a dashed line is used to represent a connection between one terminal of the element and a reference terminal for those situations where a material linkage does not exist. This would be true, for example, for the acceleration of a mass in relation to an inertial reference, for the temperature of a thermal capacitor relative to absolute zero, or for gravitational force acting through a distance.

The power or rate at which energy is flowing into the mass (Figure 9-8) is given by

$$\mathbf{P}_m = f_m v_m \qquad (9\text{-}7)$$

Figure 9-8 Conceptual model of a translational mass.

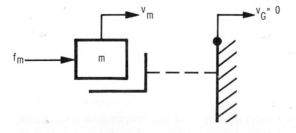

[8] The theory of relativity has shown that in general there is a nonlinear relationship between linear momentum and velocity. However, in nearly all engineering applications the assumption of an ideal mass is valid.

The force, f_m, is transmitted *through* the mass, and the velocity, v_m, is measured *across* the mass (with respect to a fixed frame). If energy is flowing into the mass there must be some way of *storing* this energy. Energy which is stored by a mass due to its motion is called its kinetic energy and is given by the triangular area shown in Figure 9-7. Since the area of a triangle is $\frac{1}{2}$(base)(height), the energy stored by a mass that has a linear momentum, p_m, and a velocity, v_m, is

$$\text{Energy} = \text{area} = \frac{1}{2}(p_m)(v_m)$$

Using the expression for an ideal mass ($p_m = mv_m$), we have

$$E_m = \frac{1}{2}mv_m^2 \tag{9-8}$$

Because the energy stored by the mass is a function of its across variable,[9] velocity, we can categorize it as an *across variable energy storage element.*

Example A mass of 2 kilograms is subjected to a force of 5 newtons. Calculate the time rate of change of the mass' velocity.

Solution

$$a = \frac{dv_m}{dt} = \frac{1}{m}f_m = \frac{5 \text{ newtons}}{2 \text{ kilograms}} = \textbf{2.5 meters/sec}^2$$

Summarizing the properties of an ideal translational mass:

Elemental equation $\qquad\qquad \dfrac{dv_m}{dt} = \dfrac{1}{m}f_m$

Energy stored $\qquad\qquad\qquad E_m = \frac{1}{2}mv_m^2$

Conceptual model

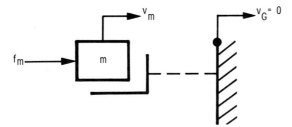

Figure 9-9

Problem

9-6. A rocket has a mass of 400,000 kg and is traveling in outer space. The rocket's thrustors have a thrust capability of 40,000 newtons. If they are turned on, calculate the acceleration, dv/dt, of the rocket.

Ideal translational spring

Springs are a common part of our daily experience. For example, scales, car suspensions, and elastic materials of all kinds exhibit the property of compressing or elongating under load. Figure 9-10 is a conceptual model of a coiled spring. An ideal

---

[9] We could have just as easily expressed the energy in terms of p_m rather than $v(E_m = \frac{1}{2}p_m^2/m)$. However, p_m is neither an across nor a through variable (it is an integrated through variable).

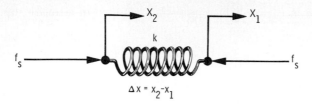

Figure 9-10 Conceptual model of a translational spring.

translational spring is defined as one that (1) has no mass, and (2) has a linear force-displacement elemental representation as shown in Figure 9-11.

In Figure 9-11, Δx represents the deflection of the spring measured from its relaxed state. The force-displacement characteristic is also not usually linear as shown in Figure 9-11 but will curve upward (hardening spring) or downward (softening spring). However, if the deflections (Δx) are small, such a linear approximation is often acceptable. The spring stiffness (k) is a function of the design of the spring and the material used in its manufacture. The flexibility of the spring is the reciprocal of its stiffness ($1/k$) and is denoted by the symbol A. Moreover, real springs have some mass; however, any real spring can be modeled as a combination of an ideal mass and an ideal spring.

The defining equation for the ideal spring is

$$f_s = k\,\Delta x = \frac{\Delta x}{A} \tag{9-9}$$

where $\Delta x = x_2 - x_1$. If we define v_s as the time rate of change of Δx, then

$$\frac{df_s}{dt} = \frac{k\,d(\Delta x)}{dt} = kv_s = \frac{v_s}{A} \tag{9-10}$$

Equation 9-10 is called the elemental equation for the spring, which is the equation that relates its across variable (v_s) to its through variable (f_s). The power or rate of change of energy of the spring is $\mathbf{P} = f_s v_s$. The energy stored in the spring can be expressed either in terms of the force, f_s, or the deflection, Δx. However, since we have chosen to use power variables to describe our ideal elements, we will use f_s. The expression for the stored energy in terms of the *through* variable f_s is given by the

Figure 9-11 Force-displacement relationship for an ideal translational spring.

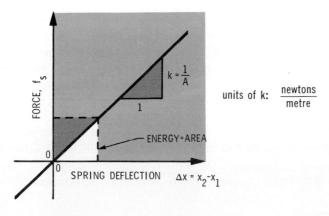

triangular area shown in Figure 9-11. Expressing this area in terms of the force, f_s, we have

$$E_s = \frac{1}{2}\frac{f_s^2}{k} = \frac{1}{2}Af_s^2 \qquad (9\text{-}11)$$

Thus we can categorize an ideal spring as a *through variable energy storage element*.

Again we note that the stored energy in a spring could be expressed in terms of its displacement; however, displacement is neither an across nor a through variable (it is an integrated[10] across variable).

Example A spring whose spring constant is $k = 100$ newtons/meter is supporting a weight of 200 newtons. Find the energy stored in the spring.

Solution

$$E_s = \frac{f_s^2}{2k} = \frac{(200 \text{ N})^2}{2(100)\text{N/m}} = \textbf{200 newton-meters} \quad (\text{joules})$$

Summarizing the properties of an ideal translational spring:

Elemental equation $\qquad\qquad \dfrac{df_s}{dt} = kv_s = \dfrac{v_s}{A}$

Energy stored $\qquad\qquad\qquad E_s = \dfrac{f_s^2}{2k} = \dfrac{1}{2}Af_s^2$

Conceptual model[11]

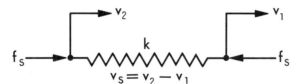

$$v_s = v_2 - v_1 \qquad\qquad \textbf{Figure 9-12}$$

Problem

9-7. An automobile suspension spring is supporting 4448 newtons and is compressed 0.025 m. Find the spring constant, k, of the spring and the energy stored in the spring.

Ideal translational damper

The ideal mass and ideal spring comprise two energy storage elements in our modeling "arsenal." It is now time to add an ideal element that is capable of dissipating energy, the *ideal damper*. The common shock absorber used in automobiles is a good example of an element that dissipates rather than stores energy. If energy is added to a spring by compressing it, the spring can return energy to the system by returning to its original position. However, if a shock absorber is compressed, it cannot return energy to the system. An ideal force–velocity relationship for an ideal damper is shown in Figure 9-13 where v_b is the relative velocity across the damper. A conceptual model of this system is shown in Figure 9-14. We assume in modeling the ideal

[10] If $v_s = d\,\Delta x/dt$, then Δx is the integral of v_s.
[11] The conceptual model is shown using through and across variables; thus v_2 and v_1 have replaced x_2 and x_1 of Figure 9-10.

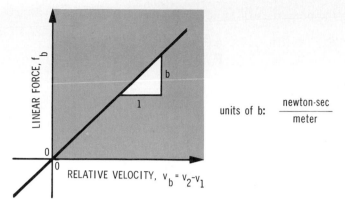

Figure 9-13   Force-velocity relationship for an ideal translational damper.

units of b: $\dfrac{\text{newton-sec}}{\text{meter}}$

damper that (1) it has no mass or elastic properties, and (2) it has a linear force-velocity relationship. If this idealized model is not adequate, it can be improved by the addition of an ideal mass and spring to model the physical damper.[12] The second assumption is valid if the damping mechanism is viscous, and if the velocity difference across the damper is relatively small. The elemental equation for the ideal damper is

$$f_b = b(v_2 - v_1) = bv_b \tag{9-12}$$

The energy flow rate or power into the damper is

$$\mathbf{P} = f_b v_b = bv_b^2 \geq 0 \tag{9-13}$$

Note that this expression is always positive (for nonzero velocity), which means that energy is *always being dissipated*. The energy is dissipated as thermal energy, causing a temperature rise in the damper.

Example   The damper shown in Figure 9-14 has its right end fixed ($v_1 = 0$) and its left end has a velocity of 10 m/s. The damping coefficient, b, has a value of 20 newton-seconds/meter. Find the force, f_b.

Solution

$$f_b = b(v_2 - v_1) = \left(20\,\frac{\text{N-s}}{\text{m}}\right)(10 - 0)\frac{\text{m}}{\text{s}}$$

$$= \textbf{200 newtons}$$

[12] A realistic mechanical damper will be modeled later as a combination of ideal elements.

Figure 9-14   Conceptual model of an ideal translational damper.

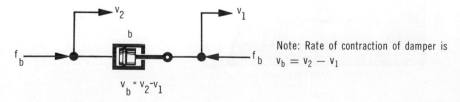

Summarizing the properties of an ideal translational damper:

Elemental equation $\qquad\qquad\qquad f_b = bv_b$

Power dissipated $\qquad\qquad\qquad \mathbf{P} = bv_b^2$

Conceptual model

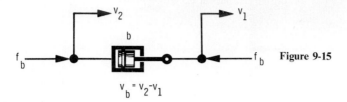

Figure 9-15

$$v_b = v_2 - v_1$$

Problem

9-8. An automobile shock absorber is subjected to a force of 1000 N and has a relative velocity across it of 0.5 m/s. Find the damping coefficient, b, and the power dissipated in watts.

Ideal sources

To complete our "arsenal" of modeling tools for mechanical translational systems, we need to define the concept of an ideal source. Since we are using power variables we will need both an ideal *through* variable source (force source) and ideal *across* variable source (velocity source). An ideal source is defined as a device that is capable of supplying a prescribed variable level (either constant or time varying) that is independent of other system variables. Theoretically this is not possible; however, in practice, it can be a useful and very nearly correct assumption. For example, the force on a mass near the surface of the earth due to gravity is very nearly constant, independent of its position. The voltage across a car battery is very nearly constant for a small level of current drawn from it. The pressure at the bottom of a deep reservoir of water is approximately constant for small changes of depth, independent of the volume flow rate out of the reservoir. Figure 9-16 illustrates a typical relationship that could be approximated as an ideal source for small changes in a system parameter. If we expect the system variable to have values outside of the valid range

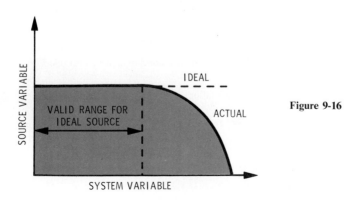

Figure 9-16

for an ideal source, we can usually model the device as a combination of an ideal source and an ideal element.

Thus, in the examples cited, gravity can be modeled as a *force source*, a car battery as a voltage source, and a reservoir as a pressure source. Other examples of ideal sources are the following:

Ice (temperature source)
Atmosphere (temperature source)
Furnace (heat flow, or more precisely *thermal energy*, source)
Motor (velocity source)
Pump (volume flow rate source)

The schematic respresentation that is used for a mechanical ideal source is shown in Figure 9-17. The sign convention used for the velocity source is indicated by the + and − signs on the circle; thus the velocity at the top of Figure 9-17a is greater than at the bottom. The sign convention for the force source shown in Figure 9-17b is to point the arrow in the direction of the applied force. These two sign conventions will be illustrated by examples.

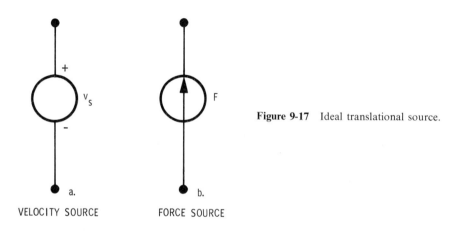

Figure 9-17 Ideal translational source.

VELOCITY SOURCE   FORCE SOURCE

Example   The spring-mass system shown in Figure 9-18 exists in a gravity field. The notation x represents the displacement of the spring from its undeflected position. Figure 9-19 shows the system using ideal elements and sources. Note that the direction of the force source to model gravity is upward since it tends to *increase* the velocity of the mass with respect to the reference velocity ($v_g = 0$).

Example: Typical Shock Absorber   Figure 9-20 is a sketch of a typical automobile shock absorber. In order to model this *dynamic system,* we will need most of the ideal mechanical elements that have been introduced. Both the outside casing and the plunger can be modeled as ideal masses. The spring connecting the plunger to the casing can be modeled as an ideal spring, and the hole in the plunger to allow fluid passage can be modeled as an ideal viscous damper.

Depending on how the shock absorber is to be used, we can have a variety of different dynamic models. First, suppose that the case mass is attached to a fixed reference and that the plunger mass is subjected to an applied force, F, that is directed

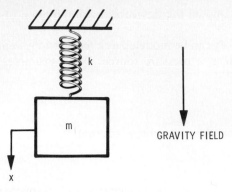

GRAVITY FIELD

Figure 9-18 Pictorial representation of a simple spring-mass system.

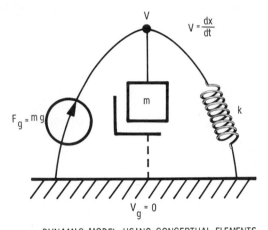

$$V = \frac{dx}{dt}$$

$F_g = mg$

$V_g = 0$

DYNAMIC MODEL USING CONCEPTUAL ELEMENTS

Figure 9-19 Conceptual model of a simple spring-mass system using ideal elements.

Figure 9-20 Typical shock absorber.

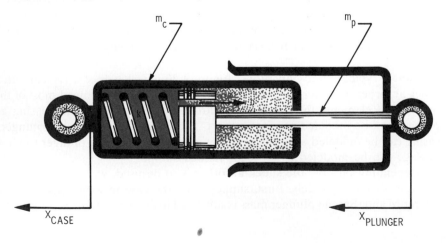

X_{CASE}

$X_{PLUNGER}$

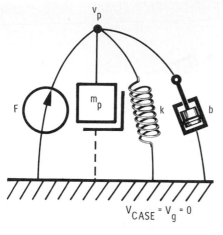

Figure 9-21 Dynamic model of shock absorber with case fixed.

in the x_p direction. Figure 9-21 is a schematic representation of this situation using ideal elements. Note that the force is modeled as an ideal source with the arrow pointing upward since the force tends to increase the velocity, v_p. It is also important to note that the ideal mass, spring, and damper are connected in parallel since they all share the same across variable, v_p.

Next, suppose that we wish to *analyze* how the shock absorber will perform when it is installed in an automotive suspension system. To do this we will assume that the plunger is rigidly connected to a quarter of the automobile's mass since there are four suspension groups in an automobile. Next, we will assume that the case is rigidly attached to an axle which is supported by a rubber tire, as shown in Figure 9-22.

We will model the tire as an ideal spring and the road's vertical irregularities as a known function of time; thus,

$$x_{\text{road}} = f(t)$$

or

$$v_{\text{road}} = \frac{dx_{\text{road}}}{dt} = \frac{df(t)}{dt}$$

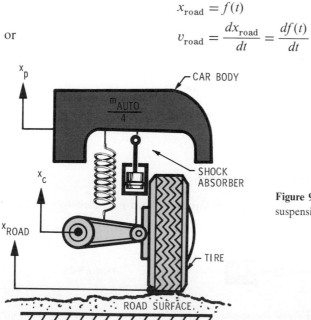

Figure 9-22 Schematic of an automobile suspension.

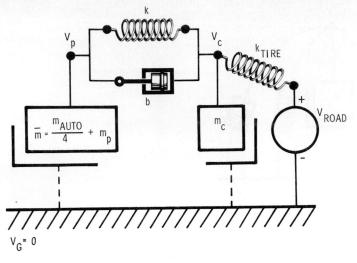

Figure 9-23 Dynamic model of an automobile suspension.

A dynamic model of the automobile suspension system using ideal elements[13] is shown in Figure 9-23.

It is important to note that the effect of gravity has not been included in this model; it could be included by the addition of two force sources on the two masses as was done in Figure 9-19. Also, note the sign of the velocity source; we are defining v_{road} to be positive with respect to our reference velocity, $v_G = 0$.

After we present the remaining fundamental building blocks for all the different media, we will develop a rational procedure to formulate the mathematical *equations* that predict the behavior of a system, such as that shown in Figure 9-23.

Problems

9-9. A constant force is applied to a mass as shown in Figure 9-24. The *dynamic model* for this system is composed of the *conceptual* models of a force source and a mass. Draw the dynamic model of this system.

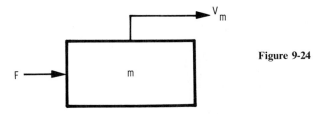

Figure 9-24

9-10. A constant force is applied to a damper as shown in Figure 9-25. Sketch the dynamic model.

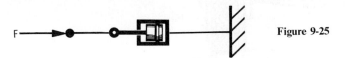

Figure 9-25

[13] Since the plunger mass and the automobile mass are rigidly attached, we can combine them as $\bar{m} = \dfrac{m_{auto}}{4} + m_p.$

9-11. A constant force is applied to the mass damper system shown in Figure 9-26. Sketch the dynamic model.

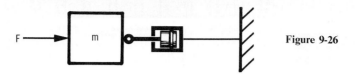

Figure 9-26

9-12. A constant force is applied to the mass-spring-damper system shown in Figure 9-27. Sketch the dynamic model.

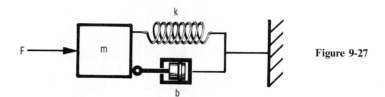

Figure 9-27

9-13. Figure 9-28 shows a simple vibration isolator. It is desired to keep the machine mass, m_m, from vibrating too severely when the floor is subject to a vibratory velocity input, $v_f(t)$. The attached spring-mass system is designed so that the combination of the mass, m_a, and the spring, k_a, will *absorb* the energy while the machine mass, m_m, remains relatively motionless. Sketch the dynamic model of the system.

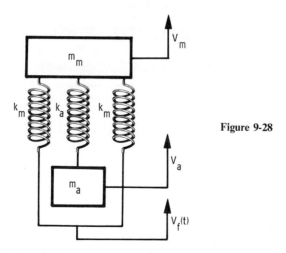

Figure 9-28

9-14. Assume in Problem 9-13 that the base of the machine is fixed ($v_f = 0$) and that the machine mass, m_m, is subject to a force $F(t)$ which is in the direction of v_m. Sketch the dynamic model.

9-15. Vibration isolators (also called vibration absorbers) are sometimes installed on the top floor of large skyscrapers. The purpose of these devices is to absorb the energy of gusting winds and thus to relieve the load on the building's structural elements. Figure 9-29 shows a pictorial representation of one of these systems. Assume that the wind imparts a velocity $v_b(t)$ to the top floor. Sketch the dynamic model of the vibration absorber assuming $v_b(t)$ to be a velocity source.

9-16. Another important consideration in the design of large skyscrapers is their ability to withstand the effect of earth tremors or earthquakes. The building's walls have a certain

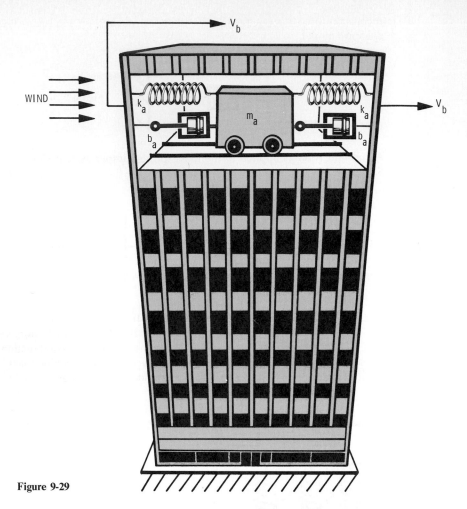

Figure 9-29

stiffness and damping. A very simple pictorial model of the building's resistance to a side to side velocity caused by an earth tremor is shown in Figure 9-30. Sketch the dynamic model.

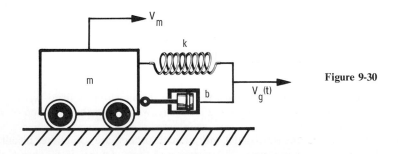

Figure 9-30

9-17. A simplified model of a vehicle suspension system is shown in Figure 9-31. The passenger compartment has a mass, m_p; the secondary suspension system is composed of a spring, k_s, and a damper, b_s; the undercarriage mass is m_u; and the tire has a spring constant, k_t. The passenger compartment is subjected to a force, $F(t)$, and the ground excitation is modeled by a velocity source, $v_g(t)$. Sketch the dynamic model (neglect the influence of gravity).

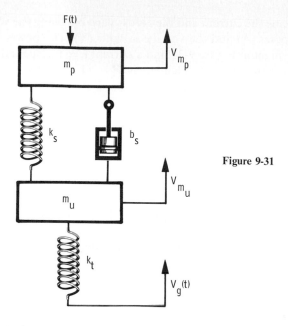

Figure 9-31

9-18. Include the effect of gravity in Problem 9-17 by modeling the gravity force on each mass by a force source.

Electrical systems

In this section we analyze electrical systems and model them by constructing simple ideal elements, much in the same way that we considered mechanical systems. In fact, the analogies between mechanical and electrical systems should become clear when we see the similarities between the functions of *energy storage and dissipation*.

The electrical power delivered to an element can be computed by multiplying the current passing *through* the element by the voltage drop *across* it, as shown in Figure 9-32; that is, $\mathbf{P} = i(e_2 - e_1)$. The *voltage* (potential difference) between points 1 and 2 is defined as the work which would be done in moving a positive charge from point 2 to point 1. The *current* is defined as the *rate of flow* of positive charge across a given area during a unit time.

We will use power variables to describe the ideal electrical elements, with the

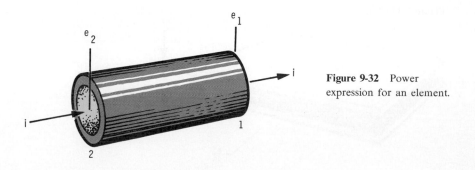

Figure 9-32   Power expression for an element.

through variable being the current and the across variable being the voltage. Current is measured in amperes (A) and the voltage in volts (V). The power is expressed in watts (W). If the element in Figure 9-32 had a current $i = 1$ amp, and a voltage drop $e = 1$ volt, then the power *into* the element is $\mathbf{P} = (1\text{ amp})(1\text{ volt}) = 1$ watt.

Ideal capacitance

Charged particles can be stored on a conducting material if two pieces of the conducting material are separated by a medium that does not allow charge to flow through it. An element that has the ability to store charge is a capacitor. Figure 9-33 shows two sheets of conducting material separated by a different medium that act as a capacitor. This ability to store charge is called *capacitance*. The physical arrangement shown in Figure 9-33 is not the only useful geometric arrangement; other possibilities exist, such as concentric spheres.

An *ideal* capacitance is one in which the charge that is stored on the capacitance is linearly proportional to the voltage between the two plates. Figure 9-34 illustrates this linear relationship, which can be expressed as

$$q_c = C(e_2 - e_1) = Ce_c \tag{9-14}$$

where e_c is the voltage across the capacitance. The constant C is called the capacitance and is in general a function of the geometry and material properties of the capacitor. We need to express this equation in terms of power variables. Therefore, by differentiating equation 9-14, we obtain

$$\frac{dq_c}{dt} = C\frac{de_c}{dt} \tag{9-15}$$

and since current is the rate of charge flow (dq_c/dt), this elemental equation becomes

$$i_c = C\frac{de_c}{dt} \tag{9-16}$$

The electric energy stored in the capacitor is the triangular area shown in Figure 9-34 and can be expressed as

$$E_c = \frac{Ce_c^2}{2} \tag{9-17}$$

Therefore, the energy stored is a function of the *across* variable, voltage. Thus it is an across variable storage element, just as the ideal mass stores energy by virtue of its across variable, velocity.

Figure 9-33 Simple capacitor.

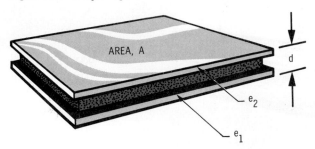

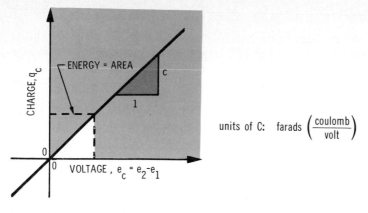

units of C: farads $\left(\dfrac{coulomb}{volt}\right)$

Figure 9-34 Charge-voltage relationship for an ideal capacitor.

Example In the capacitor shown in Figure 9-33 the voltage across the capacitor ($e_c = e_2 - e_1$) is changing with a time rate of change of 500 volts/sec. The value of the capacitance is $C = 0.50$ microfarad (0.50×10^{-6} farad). Find the current passing through the capacitance.

Solution

$$i_c = C\frac{de_c}{dt} = (0.5 \times 10^{-6}\ \text{farad})(500\ \text{volts/sec})$$

$$= \mathbf{2.5 \times 10^{-4}\ ampere}$$

Summarizing the properties of an ideal capacitor:

Elemental equation $\dfrac{de_c}{dt} = \dfrac{1}{C}i_c$

Energy stored $E_c = \dfrac{Ce_c^2}{2}$

Conceptual model

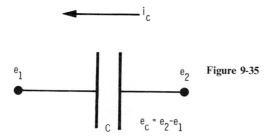

Figure 9-35

The analogy between the ideal mass and the ideal capacitance can be seen more clearly by comparing the respective elemental and energy storage equations. The analogous quantities are the mechanical velocity and the electrical voltage (across variables), the mechanical force and the electrical current (through variables), and the mechanical mass and electrical capacitance.

Problems

9-19. At a particular instant a capacitor has a capacitance $C = 1 \times 10^{-7}$ farad, and a current flowing through it of $i = 1 \times 10^{-4}$ amp. Find the rate of change of voltage across it.

9-20. At a particular instant a capacitor has a current flowing through it of $i = 1 \times 10^{-5}$ amp and a rate of change of voltage across it of $de_c/dt = 100$ volts/s. Find the capacitance of the capacitor.

9-21. The capacitor of Problem 9-19 has a voltage across it of $e_c = 0.1$ volt. Find the energy in joules stored in the capacitor and the power in watts being supplied to it.

9-22. The capacitor of Problem 9-20 has a stored energy of 1×10^{-6} joule. Find the voltage across it.

Ideal inductance

The next ideal element that we need is one that stores energy by virtue of its through variable, the current, which is the electrical equivalent of the mechanical spring. When current flows through a conductor a magnetic field is established. The energy that is stored in this field is called electromagnetic energy, and the magnitude of this stored energy depends on the magnitude of the current flowing through the element. This property of storing energy by virtue of current flow is called *inductance*.

An ideal *inductance* is defined as an element in which the voltage across the element is linearly proportional to the *rate of change of current* through it, or

$$e_2 - e_1 = e_L = L\frac{di_L}{dt} \qquad (9\text{-}18)$$

The constant L is called the inductance of the coil. Figure 9-36 is a conceptual model of an ideal inductor; Figure 9-37 shows the ideal inductor's linear characteristic.

Figure 9-36 Conceptual model of an ideal inductor.

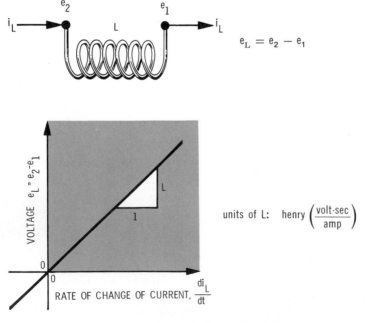

$$e_L = e_2 - e_1$$

units of L: henry $\left(\dfrac{\text{volt-sec}}{\text{amp}}\right)$

Figure 9-37 Voltage-current relationship for an ideal inductor.

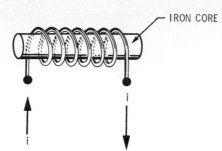

IRON CORE

Figure 9-38 Solenoid coil.

An example of a practical inductor is the solenoid coil shown in Figure 9-38. The inductance of the coil is a function of the number of turns of wire around the rod, the type of rod material, and the length of the rod.

The energy stored in an inductor is

$$E_L = \tfrac{1}{2} L i_L^2 \qquad (9\text{-}19)$$

Example The current through the inductor shown in Figure 9-37 is changing at a rate of 100 amp/sec. The inductance of the coil is $L = 2 \times 10^{-3}$ henry. Find the voltage drop across the inductor.

Solution

$$e_L = L \frac{di_L}{dt} = (2 \times 10^{-3} \text{ henry})(100 \text{ amp/sec}) = \mathbf{0.2 \ volt}$$

Summarizing the properties of the ideal inductor:

Elemental equation $\dfrac{di_L}{dt} = \dfrac{1}{.L} e_L$

Energy stored $E_L = \tfrac{1}{2} L i_L^2$

Conceptual model

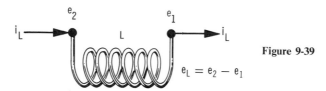

$e_L = e_2 - e_1$

Figure 9-39

By comparing the summaries of the electrical inductor with those of the mechanical spring we see that the analogous variables are again the force and current, the velocity and voltage, and the inductance L and the spring flexibility A (or reciprocal of the spring constant, $1/k$).

Problems

9-23. A coil has an inductance of $L = 1 \times 10^{-3}$ henry and a voltage across it, $e_L = 0.1$ volt. Find the rate of change of current through it.

9-24. At a particular instant a coil has a voltage across it of 1 volt and a rate of change of current through it of 200 amp/s. Find the inductance of the coil.

9-25. The coil of Problem 9-23 has a current flowing through it of $i_L = 60$ amp. Find the energy stored in the coil and the power flowing into it.

9-26. The coil of Problem 9-24 has a stored energy of 10^{-6} kilowatt-hour (or 36 joules). Find the current flowing through it.

Ideal resistance

We have formulated the energy storage elements, one for across variable energy storage (capacitors), and one for through variable energy storage (inductors). We now need an element that dissipates energy. The electrical analogy to the mechanical damper is the electrical resistor. When charged particles flow through a conductor, there is a *resistance* to this flow, called simply the resistance of the conductor. Figure 9-40 is a conceptual model of current flowing through a section of a conductor.

Figure 9-40 Conceptual model of an ideal resistance.

An ideal resistor is defined as one in which the voltage drop across it is linearly proportional to the current through it, or

$$e_R = e_2 + e_1 = Ri_R \qquad (9\text{-}20)$$

This is known as Ohm's Law; Figure 9-41 shows this linear relationship.

The power or rate of energy flow into the resistor is

$$\mathbf{P} = e_R i_R = \frac{e_R^2}{R} = i_R^2 R \geq 0$$

Note that just as in the case of the mechanical damper, the power always flows into the damper; that is, it is always dissipating.

Example The resistance shown in Figure 9-40 has a value of $R = 200$ ohms. If the voltage drop across the resistance is 50 volts, find the current through it.

Figure 9-41 Voltage-current relationship for an ideal resistor.

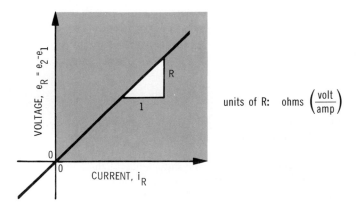

units of R: ohms $\left(\dfrac{\text{volt}}{\text{amp}}\right)$

Solution

$$i_R = \frac{e_r}{R} = \frac{50 \text{ volts}}{200 \text{ ohms}} = \textbf{0.25 amp}$$

In summarizing the properties of the ideal resistor:

Elemental equation $\qquad\qquad e_R = Ri_R$

Power dissipated $\qquad\qquad \mathbf{P} = i_R^2 R$

Conceptual model

$$e_R = e_2 - e_1$$

R

Figure 9-42

Problems

9-27. A wire of resistance $R = 100$ ohms has a current of 10 amps passing through it. Find the voltage across it and the power dissipated by it.

9-28. In order to experimentally determine the resistance of a given length of wire, a constant voltage of 6 volts was maintained across it, and the current through it was measured to be 0.01 amp. What is the resistance of the wire?

Ideal sources

Just as we needed to include ideal elements to model mechanical system *inputs,* we also need ideal electrical input models. Again, since we are using power variables, the two types of sources that we need are (1) a *through* variable source (current source) and (2) an *across* variable source (voltage source). An ideal source is one that is capable of maintaining either a prescribed through or across variable level, independent of other system variables. Figure 9-43 shows the schematic of the ideal electrical sources.

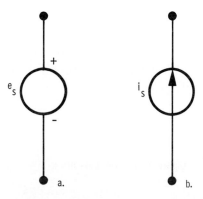

Figure 9-43 Electrical sources.

IDEAL VOLTAGE SOURCE   IDEAL CURRENT SOURCE

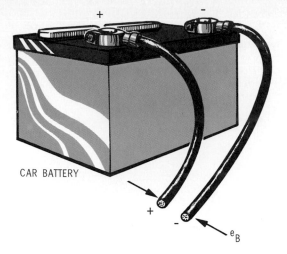

Figure 9-44 Standard 12-volt car battery.

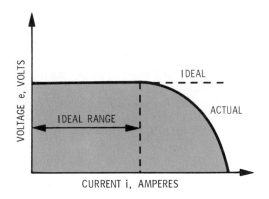

Figure 9-45 Voltage-current relationship for a standard automobile battery.

Figure 9-44 shows a standard car battery. If we model this as an ideal voltage source that supplies a constant 12 volts to any system, it must have a voltage-current relation such as that shown for the ideal range in Figure 9-45. In practice, however, as the system draws more and more current from the battery, the voltage drops off as shown. We will see later how the actual car battery performance shown in Figure 9-45 can be modeled as a combination of an ideal voltage source and an ideal resistor.

Example Figure 9-46 shows a simple electrical circuit which is called a "low-pass filter." The voltage, e_{in}, is a prescribed function of time; therefore, we can model it as an ideal voltage source. Figure 9-47 shows a slightly rearranged form of Figure 9-46.

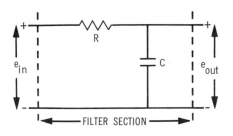

Figure 9-46 Low-pass filter.

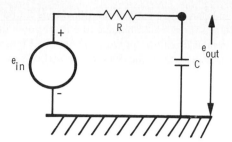

Figure 9-47 Dynamic model of a low-pass filter.

Problems

9-29. Sketch the dynamic model of a circuit composed of a voltage source that is in series with an inductor and a resistor.

9-30. Draw the dynamic model of the pictorial system shown in Figure 9-48. (*Hint:* The light bulb can be modeled as a resistance.)

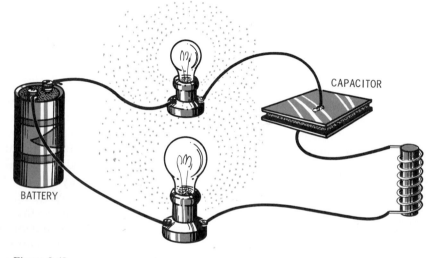

Figure 9-48

9-31. Sketch the dynamic model of the circuit shown in Figure 9-49. The symbols *R, C,* and *L* represent resistance, capacitance, and inductance.

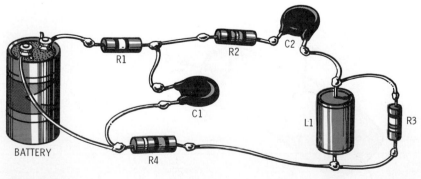

Figure 9-49

9-32. An automobile battery is tested and its voltage-current relationship is found to be like that in Figure 9-45. An ideal voltage source would behave like that shown in the ideal range. A good approximation for such a realistic curve as shown in Figure 9-45 is a voltage source in series with a resistance. Sketch the dynamic model of such a realistic battery.

9-33. Suppose the realistic automobile battery described in Problem 9-32 is connected to a resistor in series with a solenoid coil (inductor). Sketch the dynamic model of this system.

9-34. An analog computer may be composed of electrical elements that are connected in such a way that they simulate the system of interest. Suppose it is desired to simulate the mechanical system described in Problem 9-12. Sketch a dynamic model composed of *electrical* elements that would simulate the mechanical system.

9-35. Sketch the dynamic model of the electrical analog of Problem 9-13.

9-36. Sketch the dynamic model of the electrical analog of Problem 9-15.

9-37. Sketch the dynamic model of the electrical analog of Problem 9-16.

9-38. Sketch the dynamic model of the electrical analog of Problem 9-17.

Fluid systems

In this section we consider simple fluid elements that are often used in models of real systems. In particular we will consider fluids that are constrained to move in pipes or tubes or are stored in containers under pressure. The operation of machine tool equipment, movement of aircraft control surfaces, actuation of automotive power brakes, and so on, are applications of using moving fluids to do work. The study of working fluids in motion is one aspect of the field of hydraulics. In general, fluid elements tend to be more nonlinear than mechanical or electrical system elements, but valid preliminary design decisions can often be made on the basis of idealized linear behavior. In this section we consider only the ideal behavior of fluid elements.

Figure 9-50 shows a section of tubing of area, A, with a quantity of fluid passing through it. At any instant in time the volume of fluid contained in the small element of length dx is $d\text{Vol} = A\,dx$. We will assume that the pressure acting over the area A, is a constant, where pressure is defined as the force (F) on the area of fluid divided by the area (A); that is,

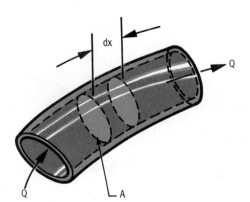

Figure 9-50 Section of tubing.

$$P = \frac{F}{A} \quad \text{or} \quad A = \frac{F}{P}$$

By substitution we have

$$\frac{F}{P}(dx) = d\mathrm{Vol} \quad \text{and} \quad F\,dx = P\,d\mathrm{Vol}$$

The volume flow rate (Q) is the quantity of fluid passing through the element (dx) in the time dt, or $Q = d\mathrm{Vol}/dt$. Now since $F\,dx = P\,d\mathrm{Vol}$, we can divide both sides of the equation by dt and obtain

$$F\frac{dx}{dt} = P\frac{d\mathrm{Vol}}{dt}$$

or

$$Fv = PQ \tag{9-21}$$

The expression Fv (force times velocity) represents the mechanical power or time rate of change of energy. The expression for fluid power is PQ or the product of pressure ($\mathrm{N/m^2}$) and volume flow rate ($\mathrm{m^3/s}$).

Following our convention of using power variables to describe system behavior we will consider pressure to be the *across variable* since pressure is measured between two points. We will consider volume flow rate as the *through variable* since it is the quantity of fluid passing *through* an element in a small period of time, dt.

Ideal elements

Our next task is to define a set of simple fluid elements so that we can model practical systems with a *conceptual model* that is composed of simple elements. Just as we did in the mechanical and electrical cases, we need to define the following.

1. An element that stores energy by virtue of its across variable, pressure.
2. An element that stores energy by virtue of its through variable, volume flow rate.
3. An element that dissipates energy.
4. An ideal across variable source (pressure source).
5. An ideal through variable source (volume flow rate source).

Ideal fluid capacitance

An ideal fluid capacitance is an element which stores energy by virtue of the pressure difference across it. For example, Figure 9-51 shows a fluid tank open to the atmospheric pressure with area, A, height of fluid, H, and volume flow rate into the tank, Q. If the fluid is modeled as incompressible,[14] the volume flow rate entering is equal to $A(dH/dt)$. The pressure difference across the tank, $P_2 - P_1$, is equal to the weight of the fluid in the tank divided by the area of the tank,[15] or

$$P_2 - P_1 = \frac{\rho g H A}{A} = \rho g H \tag{9-22}$$

[14] Most liquids can be modeled as incompressible, while gases must often be modeled as compressible.
[15] ρ is the mass density of the fluid, $\mathrm{kg/m^3}$; g is acceleration due to gravity, in this chapter assumed to be $9.8\ \mathrm{m/s^2}$.

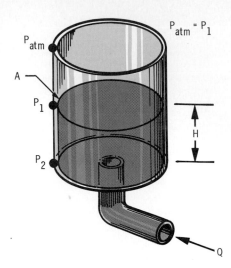

Figure 9-51   Ideal fluid capacitance.

If we adopt the nomenclature that $P_c = P_2 - P_1$ (where P_1 = constant reference pressure), we can rewrite equation 9-22 as

$$H = \frac{P_c}{\rho g} \qquad (9\text{-}23)$$

If we substitute this expression into the expression for volume flow rate, we have

$$Q = A\frac{d}{dt}\left(\frac{P_c}{\rho g}\right) = \frac{A}{\rho g}\frac{dP_c}{dt} \qquad (9\text{-}24)$$

Next we can define the quantity $A/\rho g$ (which has units of m^5/N) as a capacitance, C, and can substitute and rewrite equation 9-24 as

$$Q = C\frac{dP_c}{dt} \qquad (9\text{-}25)$$

Note the similarity of this expression with that derived for the electrical capacitance in equation 9-16. In fact, if we replaced the electrical through variable, i_c, with the fluid through variable, Q, and the electrical across variable, V_c, with the fluid across variable, P_c, we would have identical expressions.

In general, an ideal linear fluid capacitance is defined as an element that has a volume-pressure characteristic as shown in Figure 9-52. The energy stored in the tank is equal to the area under the V vs P_c curve. For the linear range, this is

$$E = \tfrac{1}{2}VP_c = \tfrac{1}{2}CP_c^2 \qquad (9\text{-}26)$$

By substituting the previously defined values of C and P_c into equation 9-26, the energy is seen to be equal to the weight of fluid, $\rho g h A$, in the tank times the height to the fluid mass center, $h/2$.

An open tank is not the only fluid element that can store energy by virtue of its pressure; others are shown in Figure 9-54. All these elements can be modeled as an ideal fluid capacitance.

Example   The tank shown in Figure 9-51 has a cross-sectional area of 1 m². The density of water is 1000 kg/m³, and the acceleration due to gravity is $g = 9.8$ m/s². Find the capacitance of the tank.

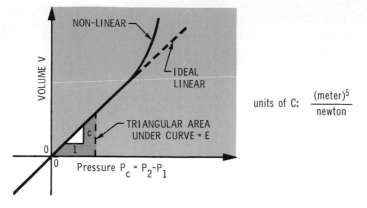

units of C: $\dfrac{(\text{meter})^5}{\text{newton}}$

Figure 9-52   Volume-pressure relationship for an ideal fluid capacitance.

Solution

$$C = \frac{A}{\rho g} = \frac{1 \text{ m}^2}{(1000 \text{ kg/m}^3)(9.8 \text{ m/s}^2)} - 1.02 \times 10^{-4} \text{ m}^5/\text{N}$$

Summarizing the properties of an ideal fluid capacitance:

Elemental equation $\dfrac{dP_c}{dt} = \dfrac{1}{C} Q_c$

Energy stored $E_c = \frac{1}{2} C P_c^2$

Conceptual model

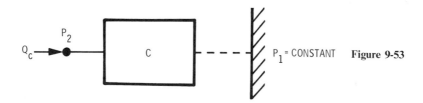

Figure 9-53

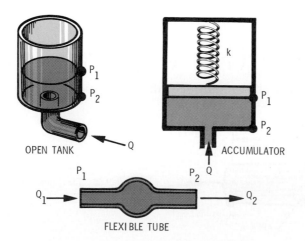

Figure 9-54   Examples of ideal fluid capacitors.

Problems

9-39. The tank shown in Figure 9-51 has a cross-sectional area of 5 m². If water is the fluid medium, calculate the fluid capacitance of the tank.

9-40. The tank of Problem 9-39 has a net volume flow rate into it of $Q_c = 100$ m³/sec. Calculate the rate of change of pressure across it. If at this particular instant the pressure across it is 9.8×10^4 N/m², find the stored energy in the tank.

9-41. A fluid capacitor has a capacitance of 1×10^{-3} m⁵/N. At a particular instant it has a pressure drop across it of 9.6×10^4 N/m². Find the energy stored in the capacitor.

9-42. A flexible tube has the property of fluid capacitance due to its elasticity. In order to determine the value of the capacitance of a certain tube, an experiment was devised. It was observed that at a particular instant the rate of change of pressure across the section of tubing was 100 N/m²-s, while the net flow through it was 6×10^{-3} m³/s. Calculate the capacitance of the tube.

Ideal fluid inertance

A fluid element which stores energy by virtue of its through variable, volume flow rate, is called fluid inertance (I). The term inertance is used because inertia forces are associated with the acceleration of the fluid in a pipe. Figure 9-55 shows a fluid in a pipe of length, ℓ, and area, A. The net force on the mass of fluid in the length, ℓ, is

$$F = A(P_2 - P_1) = AP_I$$

By applying Newton's second law and noting the fact that the velocity of the fluid is equal to the flow rate, Q, divided by the area, A, we have

$$P_I = \frac{\rho \ell}{A} \frac{dQ}{dt} \tag{9-27}$$

Again, if we define $I = \rho \ell / A$, equation 9-27 becomes

$$P_I = I \frac{dQ}{dt} \tag{9-28}$$

where I is called the fluid inertance and has units of N-s²/m⁵. Note the similarity between equations 9-28 and 9-18. The analogous quantities are voltage (pressure), current (volume flow rate), and inductance (inertance). The kinetic energy of the fluid in the pipe is equal to one half the fluid mass times the velocity squared, where the fluid velocity is Q_I/A.

$$E_I = \tfrac{1}{2} \text{ mass (velocity)}^2$$
$$= \tfrac{1}{2}(\rho A \ell)\left(\frac{Q_I}{A}\right)^2 = \tfrac{1}{2}\left(\rho \frac{\ell}{A}\right)Q_I^2 = \tfrac{1}{2}IQ_I^2 \tag{9-29}$$

The term Q_I is the fluid volume flow rate through the fluid element. The units of fluid inertance are N-s²/m⁵.

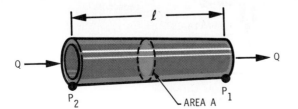

Figure 9-55 Ideal fluid inertance.

Example The pipe shown in Figure 9-55 has a length of 10 meters and an area of 0.01 m². The density of water is 1000 kg/m³. Find the inertance of the pipe if water is flowing through it.

Solution

$$I = \frac{\rho \ell}{A} = \frac{(1000 \text{ kg/m}^3)\left(\dfrac{\text{in-s}^2}{\text{m kg}}\right)(10 \text{ m})}{0.01 \text{ m}^2} = 1 \times 10^6 \text{ N-s}^2/\text{m}^5$$

Summarizing the properties of an ideal fluid inertance:

Elemental equation $\qquad\qquad\qquad \dfrac{dQ_I}{dt} = \dfrac{1}{I} P_I$

Energy stored $\qquad\qquad\qquad\quad E_I = \frac{1}{2} I Q_I^2$

Conceptual model

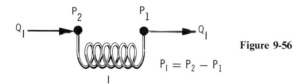

$$P_I = P_2 - P_1$$

Figure 9-56

Problems

9-43. The pipe shown in Figure 9-50 has a length of 15 m and an area of 0.002 m². If water is the fluid flowing through it, find the inertance of the pipe.

9-44. The inertance of a given section of tubing is known to be 1×10^5 N-s²/m⁵. If the pressure drop across it is 1×10^5 N/m², find the rate of change of volume flow rate through it.

9-45. Experimental measurements have determined that the pressure drop across a section of tubing is 2×10^6 N/m² while the rate of change of volume flow rate through it is 9×10^{-2} m³/s². Find the inertance of the section.

Ideal fluid resistance

There are many different mechanisms by which energy can be dissipated by a fluid in motion. Some of these include friction losses in a long tube (length of tube is much greater than the tube diameter), turbulent flow, flow through a porous medium, and flow through a sharp-edged orifice. We may recall from the previous sections on mechanical and electrical systems that the energy dissipator was defined as an element in which the through and across power variables are related by a constant ($f = bv$, $e = Ri$). Similarly, in a fluid system an ideal linear resistance will be defined as an element in which the fluid flow rate is linearly proportional to the pressure; that is,

$$P_R = RQ_R \geq 0 \qquad\qquad (9\text{-}30)$$

This relationship is illustrated in Figure 9-57.

Most fluid resistors are highly nonlinear in nature and the use of an ideal linear resistor is only valid over a small range of either pressure or flow rate. An example of

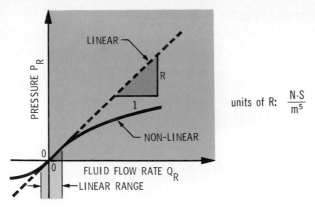

units of R: $\dfrac{\text{N·S}}{\text{m}^5}$

Figure 9-57 Pressure-flow rate relationship for a fluid resistance.

an ideal fluid resistance is the flow through a long tube. It can be shown that the flow resistance of a long tube for an incompressible fluid is:

$$R = \frac{128\mu\ell}{\pi d^4}\,\text{N-s/m}^5 \qquad (9\text{-}31)$$

where μ = absolute viscosity of the fluid, N-s/m²
  ℓ = length of tube, m
  d = diameter of tube, m

Example A long thin tube has a length of 30 m and a diameter of 1 cm. The absolute viscosity of the fluid flowing through the tube is $\mu = 0.36$ N-s/m². Find the fluid resistance of the pipe.

Solution

$$R = \frac{128\mu\ell}{\pi d^4} = \frac{128(0.36\ \text{N-s/m}^2)(30\ \text{m})}{3.14(1\times10^{-2}\ \text{m})^4} = \mathbf{4.4\times10^{10}\ \text{N-s/m}^5}$$

Summarizing the properties of an ideal fluid resistance:

Elemental equation $\qquad p_R = RQ_R$

Power dissipated $\qquad \mathbf{P} = p_R Q_R = RQ_R^2 \geq 0$

Conceptual model

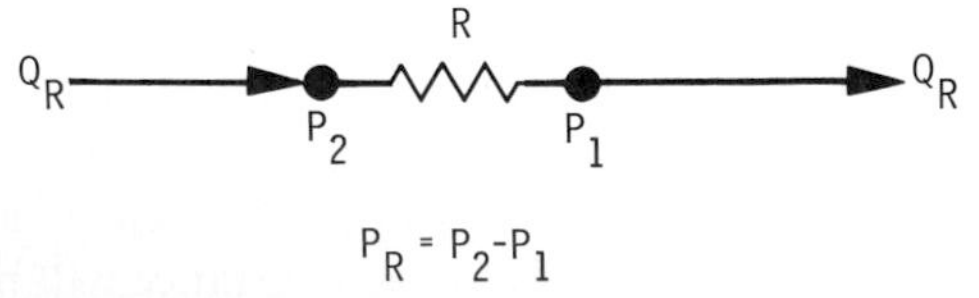

$$Q_R \xrightarrow{\hspace{2cm}} \underset{P_2}{\bullet}\!\!-\!\!\text{\Large www}\!\!-\!\!\underset{P_1}{\bullet} \xrightarrow{\hspace{2cm}} Q_R$$

$$p_R = P_2 - P_1$$

Figure 9-58

Problems

9-46. A tube has a length of 100 m and a diameter of 0.1 m. The viscosity of the fluid flowing through the tube is $\mu = 0.30$ N-s/m². Find the resistance of the pipe.

9-47. For the pipe of Problem 9-46 it is known that the volume flow rate flowing through it is constant and equal to 4×10^{-1} m³/s. Find the pressure drop across the 100-m section.

9-48. The arteries of the cardiovascular system present a resistance to fluid flow due to the viscosity of the blood. Suppose it has been experimentally determined that during a certain time interval the volume flow rate and pressure drop across a given length of artery are approximately constant and equal to $\bar{Q}$ and $\bar{p}$, respectively. Find the resistance and the power being dissipated.

Ideal sources

In this section we need to define an ideal through variable source (volume flow rate source) and an ideal across variable source (pressure source). An example of a volume flow rate source is a constant displacement pump; examples of pressure sources are the pressure at the bottom of a large reservoir and a constant pressure pump. Figure 9-59 shows the schematic representation of these ideal sources.

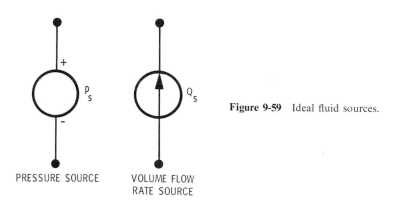

PRESSURE SOURCE   VOLUME FLOW RATE SOURCE

Figure 9-59 Ideal fluid sources.

Example: water supply system   Figure 9-60 shows a sketch of a city water supply system that is composed of a mountain reservoir, a long section of pipe, a storage tank and a constant displacement pump. Figure 9-61 shows a dynamic model of the water system. In drawing this circuit diagram we have made several important modeling assumptions. First, the reservoir has been modeled as a constant pressure source. This assumption is reasonable if the flow through the system does not appreciably affect the height of the water in the reservoir. Second, the long pipe has been modeled to

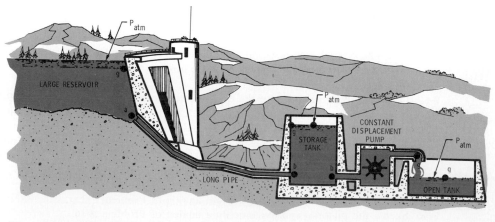

Figure 9-60 Water supply system.

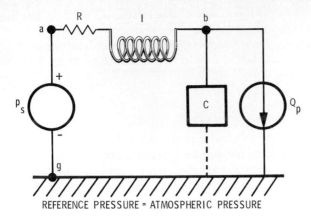

Figure 9-61 Dynamic model of a water supply system.

REFERENCE PRESSURE = ATMOSPHERIC PRESSURE

include both a fluid resistance and a fluid inertance that are connected in series since they have a common through variable (volume flow rate). The fluid capacitance of the storage tank and the constant displacement pump are connected in parallel since they have a common cross variable (pressure difference).

Problems

9-49. Sketch the dynamic model of a fluid system composed of a pressure source that is in series with a pipeline that has both resistance and inertance.

9-50. Draw the dynamic model of the fluid system shown in Figure 9-62.

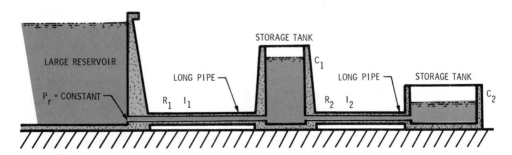

Figure 9-62 Fluid system.

9-51. Draw the dynamic model of the heart system shown in Figure 9-63.

   Note The arterial system should be modeled as a single tube that has the properties of resistance, inertance, and capacitance. The heart can be modeled as a volume flow rate source.

9-52. Draw the dynamic model of the drainage control system shown in Figure 9-64.

9-53. It is possible to build a fluidic "analog" computer using fluid elements rather than electrical elements. Such a computer would be considerably slower in response but would be less sensitive to magnetic and electrical fields encountered, for example, by missiles and aircraft. Sketch both a *pictorial* and a dynamic fluid analog model of Problem 9-12.

9-54. Sketch the pictorial and dynamic fluid analog of Problem 9-13.

9-55. Sketch the pictorial and dynamic fluid analog of Problem 9-15.

9-56. Sketch the pictorial and dynamic fluid analog of Problem 9-16.

9-57. Sketch the pictorial and dynamic fluid analog of Problem 9-17.

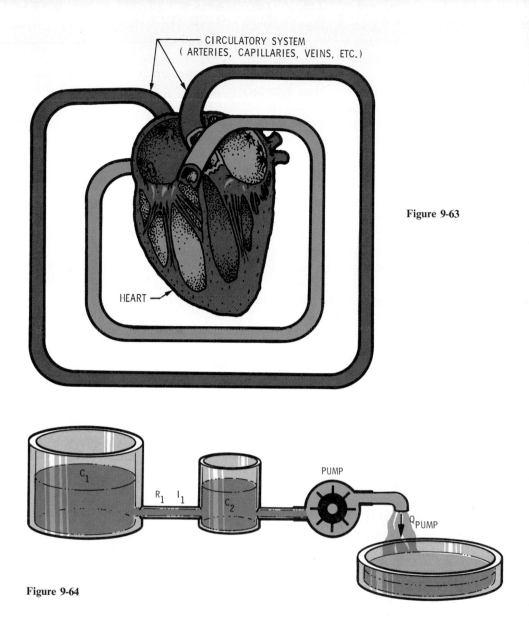

CIRCULATORY SYSTEM
(ARTERIES, CAPILLARIES, VEINS, ETC.)

Figure 9-63

HEART

PUMP

C_1

R_1 I_1

C_2

Q_{PUMP}

Figure 9-64

Thermal systems

The final types of systems that are considered in this chapter are those associated with thermal energy. The physical quantities associated with thermal energy are heat[16] and temperature. However, only simple thermal systems will be considered here, and the more complex interrelations of fluid flow, thermal energy, and mechanical motions will be reserved for study in subsequent courses. In some ways thermal systems are much like the other systems that we have considered. For example, heat flow rate (actually energy flow rate) is related to a temperature

[16] *Heat* is energy transferred from one system to another solely by reason of a temperature difference between the systems.

difference in much the same way that fluid volume flow rate is related to a pressure difference or that an electrical current flow is related to the voltage.

There are some unusual characteristics of thermal systems that make them the "odd-balls" of our systems approach to modeling. The first reason is that, in all the other types of systems that have been considered, power has been expressed as the product of a through variable and an across variable. In thermal systems, however, the power *is the through variable;* thus

$$\mathbf{P} = q \tag{9-32}$$

where q is the heat flow rate and has units of watts. The second reason for the peculiarity of thermal systems is that there is no known means of storing energy by virtue of the through variable, that is, by virtue of the heat flow through it. Thus our arsenal of ideal elements will not include a thermal "inductor," that is, a through variable energy storage element.

Our choice of variables will be the heat flow rate (through variable) and the temperature (across variable). *The units of temperature are in* °C (Celsius scale) or equivalently in K (Kelvin absolute scale).

Ideal thermal capacitance

An ideal thermal capacitance is defined as an element which stores energy by virtue of its across variable, temperature. The heat (flow rate), q, is the time rate of change of the thermal energy; that is, $q = dE_T/dt$, where E_T is the thermal energy in joules. Figure 9-65 shows the relationship for an ideal thermal capacitor where T_c represents the temperature drop across the capacitor. Thus, for an ideal linear thermal capacitance,

$$E_T = C_T T_C \tag{9-33}$$

where C_T is the thermal capacitance joule/K and T_C is the temperature in K of the element with respect to a fixed, constant temperature.

In order to express this *elemental equation* for the ideal thermal capacitor in terms of our through and across variables (q, T), we can differentiate equation 9-33, as follows:

$$\frac{dE_T}{dt} = q_c = C_T \frac{dT_c}{dt} \tag{9-34}$$

Figure 9-65 Thermal energy-temperature difference relationship for an ideal thermal capacitance.

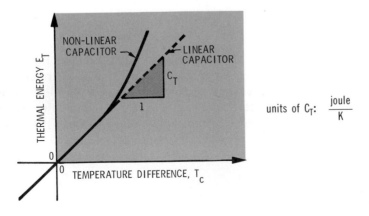

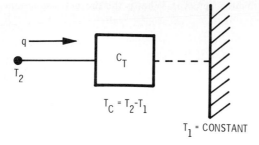

Figure 9-66 Conceptual model of a thermal capacitor.

$T_c = T_2 - T_1$

$T_1 = CONSTANT$

where q_c is the heat flow through the capacitor and T_c is the temperature of the capacitor with respect to a constant reference temperature. The thermal capacitance, C_T, is the product of the mass of the element and its specific heat,[17] C_p; that is, $C_T = mC_p$. Figure 9-66 is a conceptual model of a thermal capacitor. Our homes are full of thermal capacitors, such as drapes, couches, windows, and metallic objects.

Example The thermal capacitance of a metal block is known to be 1000 joules/K. If the instantaneous heat flow rate through the block is 300 watts, find the rate of change of temperature across it.

Solution

$$\frac{dT_c}{dt} = \frac{1}{C_T} q_c = \frac{300 \text{ joules/s}}{1000 \text{ joules/K}} = \textbf{0.3 K/s}$$

Summarizing the properties of an ideal thermal capacitance:

Elemental equation $\qquad\qquad q_c = C_T \dfrac{dT_c}{dt}$

Energy stored $\qquad\qquad\qquad E = C_T T_c$

Conceptual model

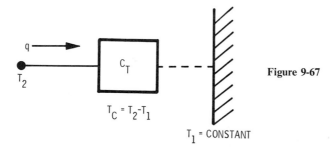

$T_c = T_2 - T_1$

$T_1 = CONSTANT$

Figure 9-67

Problems

9-58. A metal block has a constant heat flow rate through it of 500 watts. The thermal capacitance is known to be 500 joules/°K. Find the rate of change of temperature across it.

9-59. A solid object is subjected to a heat flow rate of 60 watts, and a rate of change of

[17]The specific heat of a given material is a function of many factors.

temperature of 0.05 K/s is measured across it. What is the thermal capacitance of the object?

Ideal thermal resistance

The property of a material to oppose the flow of heat (thermal energy) through it is called its thermal resistance. In general there is a temperature drop in the direction of the heat flow through it. Materials that offer little resistance to heat flow are called heat conductors (metals) and materials which offer large resistance are called insulators (fiberglass, wood).

An ideal thermal resistance is defined as one in which the heat flow, q, is linearly proportional to the temperature difference across it. Figure 9-68 shows this linear relationship. Thus the elemental equation for an ideal resistance is

$$q_r = \frac{T_r}{R_t} \tag{9-35}$$

where R_t is the thermal resistance and has units of K/watt.

There are many ways thermal energy can be transferred from one element to another. These basic mechanisms of heat transfer are conduction, convection, and radiation.

Heat conduction occurs through a material on a microscopic level from atom to atom. Figure 9-69 shows an element of length ℓ and area A through which heat conduction is occurring. The relationship between heat flow and temperature difference (reference 1) can be expressed as

$$q = \frac{\sigma A}{\ell}(T_2 - T_1)$$

Thus the thermal resistance is $R_t = \ell/\sigma A$, where σ is the thermal conductivity of the material, which can be found in many textbooks.

Heat convection is the transport of energy by movement of a mass after heat has been transferred to the mass, for example, the movement of hot air or water. A very useful concept to help analyze systems in which convection is occurring is that of a heat transfer coefficient, C_h, where

$$C_h = \frac{q}{(T_2 - T_1)A}$$

Figure 9-68 Heat-temperature differential relationship for an ideal thermal resistor.

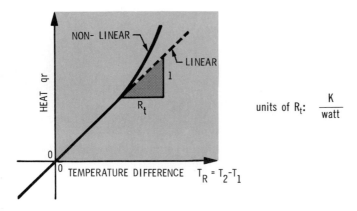

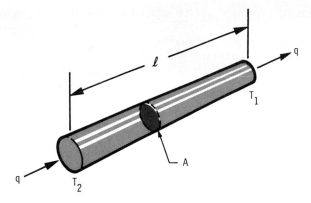

Figure 9-69   Heat conduction through an elemental length.

and has units of watt/m^2-K. The heat transfer coefficient for a given arrangement of elements and materials is often determined experimentally. From our definition of thermal resistance we see that $R_t = 1/C_h A$.

Heat radiation is a very nonlinear phenomena; in fact, all thermal energy transfer mechanisms are very nonlinear and complex. However, for *small differences in temperatures* they often can be modeled as linear ideal thermal resistors. Often in practice an approximate value is obtained *experimentally* by measuring the heat flow and temperature difference; thus

$$R_t \simeq \frac{\text{temperature difference}}{\text{heat flow rate}}$$

Example   Air at 327 K flows over a plate of area 0.23 m^2 that is maintained at 333 K. The heat transfer coefficient has been measured experimentally and has been found to be $C_h = 372.5$ watt/K(m^2). Find the heat flow rate from the plate.

Solution

$$R_t = \frac{1}{C_h A} = \frac{1}{(372.5)(0.23)} = 0.01167 \frac{K}{\text{watt}}$$

$$q = \frac{T_2 - T_1}{R_t} = \textbf{514 watts}$$

Summarizing the properties of an ideal thermal resistance:

Elemental equation
$$q_r = \frac{T_r}{R_t}$$

Conceptual model

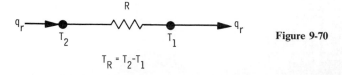

Figure 9-70

Problems

9-60. Air at 400°K flows over a plate of area 0.5 m^2 that is maintained at 425°K. The heat transfer coefficient is known to be $C_h = 400 \text{ watt/K(m}^2)$. Find the heat flow rate from the plate.

9-61. It is desired to measure experimentally the thermal resistance of a solid object. The material is subjected to a constant heat flow rate of 100 watts, and a constant temperature difference of 0.1 K is measured across it. What is its thermal resistance?

9-62. A house is maintained at a constant temperature of $21°C = 294$ K while the outside temperature is $0°C = 273$ K. It is known that the furnace is providing a constant heat flow rate of 100 watts. Find the effective thermal resistance of the combined house walls and windows.

Ideal sources

The ideal sources that we will need to complete our arsenal of thermal elements are the ideal heat flow rate source (through source) and the ideal temperature source (across source). A furnace could be modeled as a heat flow rate source, while the atmosphere could be modeled as a temperature source. It must be remembered that ideal sources do not have to be constant but they do have to be independent of the system variables. For example, if we model the atmosphere as a temperature source and a house as the system, we are assuming that the temperature changes of the house will not influence the atmospheric temperature. Schematic representations for the ideal heat flow source and temperature source are shown in Figure 9-71.

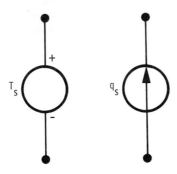

Figure 9-71   Ideal thermal sources.

Example: home heating system   The house shown in Figure 9-72 has been modeled by a thermal capacitance (walls, furniture, draperies, etc.), a thermal resistance (windows, doors), an ideal heat flow source (furnace), and an ideal temperature source (atmosphere). Of course, the walls, furniture, and the like, have some thermal

Figure 9-72   Home heating system.

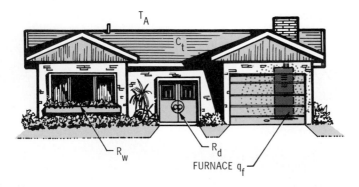

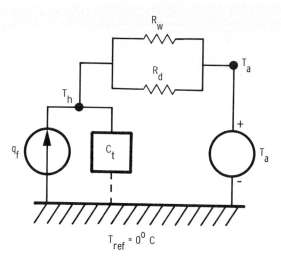

Figure 9-73 Dynamic model of a home heating system.

resistance and the windows and doors have some thermal capacitance. However, it is reasonable to analyze the problem by lumping all the capacitance into certain elements and all the resistance into others.

Figure 9-73 shows the dynamic model of the home heating system using ideal thermal elements. Notice that the furnace and house capacitance are connected in parallel since they have the same across variable (T_h, temperature of the inside of the house), and similarly that the two resistances are connected in parallel since they also have the same across variable ($T_h - T_A$).

Problems

9-63. Sketch the dynamic model of a heat flow rate source that is in series with a thermal resistance.

9-64. Sketch the dynamic model of a temperature source, a resistance, and a capacitance that are in parallel.

9-65. Draw the dynamic model of the thermal system shown in Figure 9-74. The metal has a capacitance, C_t, and is initially at a very high temperature. The surrounding insulating material has a thermal resistance of R, and the atmosphere remains at a constant temperature T_A.

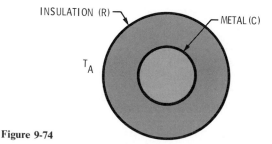

Figure 9-74

9-66. Draw the dynamic model of the thermal system shown in Figure 9-75. The metal ingot has a thermal capacitance, c; the bath solution has a thermal resistance, R_1; the liner has a thermal resistance, R_2; the atmospheric temperature remains constant at T_A.

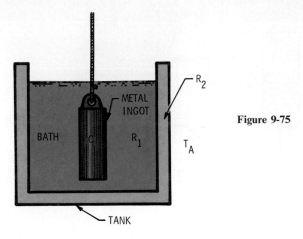

Figure 9-75

9-67. The figure shown in Figure 9-76 is a mechanical system. Sketch the dynamic model of the thermal analog of this system.

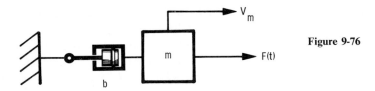

Figure 9-76

9-68. Sketch the dynamic model of the thermal analog of the electrical system shown in Figure 9-77.

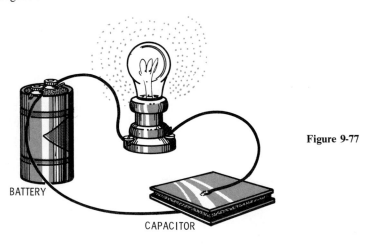

Figure 9-77

9-69. The figure shown in Figure 9-78 is a mechanical spring-mass system. Can a thermal analog of this system be drawn? Why or why not?

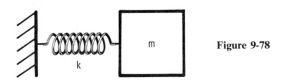

Figure 9-78

System analogies

We have just completed an elementary study of mechanical, electrical, fluid, and thermal systems. In all systems we found that the concept of through and across variables was useful in describing the behavior of ideal elements. In order to emphasize these similarities, we can define a generalized through variable (f) and a generalized across variable (V). In all cases except for thermal systems the power flow can be represented as

$$\mathbf{P} = fV \tag{9-36}$$

It is also useful to define a generalized capacitance (C), a generalized inductance (I), and a generalized resistance (R).

In all the systems we have elements that store energy by virtue of the across variable. This stored energy can be represented by (except thermal)

$$E_A = \tfrac{1}{2}CV^2 \tag{9-37}$$

We also have elements that store energy by virtue of their through variable. This stored energy can be represented by

$$E_T = \tfrac{1}{2}If^2 \tag{9-38}$$

In addition we have elements that dissipate energy at all times. The rate of change of energy (power) dissipated by these elements can be represented by

$$\mathbf{P} = Rf^2 = \frac{1}{R}V^2 \geq 0 \tag{9-39}$$

Finally we have ideal sources that maintain specified values of either the through variable or across variable independent of other system variables.

The various dynamic diagrams that were developed for all the examples can be greatly simplified by using unified schematic representations called *linear graphs*. For example, the conceptual model for the mass of a mechanical system is shown in Figure 9-79a. The corresponding linear graph is shown in Figure 9-79b. The direction of the arrow in Figure 9-79b shows that the velocity of v_m is positive with respect to the reference velocity, v_G.

Figure 9-80 illustrates how the through, across, and dissipative elements, as well as the ideal sources, can be represented by simplified schematic diagrams consisting of directed line segments.[18] There is an associated through and across variable with each element of Figure 9-80. The through variable flows from point 2 to point 1, and the across variable is measured across the two *terminals*. For example, in Figure 9-80a, $v_c = v_2 - v_1$. These concepts will be illustrated later by a number of examples. The simplified schematic diagrams of Figure 9-80 are called *linear graphs* because they are composed of a simple line with an arrow on it.

Table 9-1 summarizes all the ideal elements that we have covered so far. This table contains an abundant amount of information and points out the analogies that exist between the various types of systems, as well as places all the elemental

---

[18] In general the linear graph elements are drawn slightly curved so that a number of elements can be easily connected together.

Table 9-1 *Summary of Ideal Elements**

		Mechanical systems	Electrical systems	Fluid systems	Thermal systems	Generalized system†
Across variable energy storage elements	Conceptual element	Ideal mass	Ideal electrical capacitance	Ideal fluid capacitance	Ideal thermal capacitance	Ideal capacitance
	Conceptual model	(diagram) $V_G = 0$	(diagram)	(diagram) $P_1 = \text{CONSTANT}$	(diagram) $T_1 = \text{CONSTANT}$	(diagram)
	Linear graph	(diagram) $V_G = 0$	(diagram)	(diagram) $P_1 = \text{CONSTANT}$	(diagram) $T_1 = \text{CONSTANT}$	(diagram)
	Elemental equation	$\dfrac{dV_m}{dt} = \dfrac{1}{m} f_m$	$\dfrac{de_c}{dt} = \dfrac{1}{C} i_c$	$\dfrac{dP_c}{dt} = \dfrac{1}{C} Q_c$	$\dfrac{dT_c}{dt} = \dfrac{1}{C_t} q_c$	$\dfrac{dV_c}{dt} = \dfrac{1}{C} f_c$
	Energy stored	$E_m = \dfrac{1}{2} m V_m^2$	$E_c = \dfrac{1}{2} C e_c^2$	$E_c = \dfrac{1}{2} C P_c^2$	$E_c = C_t T_c$	$E_c = \dfrac{1}{2} C V_c^2$
Through variable energy storage elements	Conceptual element	Ideal spring	Ideal inductance	Ideal inertance	—	Ideal inductance
	Conceptual model	(diagram)	(diagram)	(diagram)	—	(diagram)
	Linear graph	(diagram)	(diagram)	(diagram)	—	(diagram)

	Ideal damper	Ideal electrical resistance	Ideal fluid resistance	Ideal thermal resistance	Ideal resistance
Elemental equation	$\dfrac{df_s}{dt} = \dfrac{1}{A}V_s = kV_s$	$\dfrac{di_L}{dt} = \dfrac{1}{L}e_L$	$\dfrac{dQ_I}{dt} = \dfrac{1}{I}P_I$	—	$\dfrac{df_I}{dt} = \dfrac{1}{I}V_I$
Energy stored	$E_S = \dfrac{1}{2}Af_s^2 = \dfrac{f_s^2}{2K}$	$E_L = \dfrac{1}{2}Li_L^2$	$E_I = \dfrac{1}{2}IQ_I^2$	—	$E_I = \dfrac{1}{2}If_I^2$
Conceptual element	Ideal damper	Ideal electrical resistance	Ideal fluid resistance	Ideal thermal resistance	Ideal resistance
Conceptual model	$V_1,\ V_2,\ f_b$	$i_R,\ e_1,\ e_2,\ R$	$Q_R,\ P_1,\ P_2,\ R$	$q_r,\ T_1,\ T_2,\ R_t$	$f_R,\ V_1,\ V_2,\ R$
Linear graph	$b,\ f_b,\ V_1,\ V_2$	$R,\ i_R,\ e_1,\ e_2$	$R,\ Q_R,\ P_1,\ P_2$	$R_t,\ q_R,\ T_1,\ T_2$	$R,\ f_R,\ V_1,\ V_2$
Elemental equation	$f_b = bv_b$	$e_R = Ri_R$	$P_R = RQ_r$	$T_R = R_t q_R$	$V_R = Rf_R$
Power dissipated	$P = bv_b^2$	$P = Ri_R^2$	$P = RQ_R^2$	$P = q_R$	$P = Rf_R^2$
Across variable sources — Linear graph	V_S	e_S	P_S	T_S	V_S
Through variable sources — Linear graph	F_S	i_S	Q_S	q_S	f_S

Energy dissipators

* Adapted from J. L. Shearer, A. T. Murphy, and H. H. Richardson, *Introduction to System Dynamics*, Addison-Wesley, 1967.
† Excluding thermal systems.

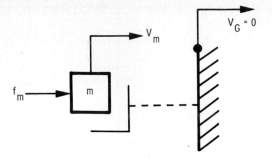

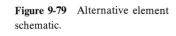

(a) CONCEPTUAL DIAGRAM OF A TRANSLATIONAL MASS

Figure 9-79 Alternative element schematic.

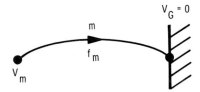

(b) SIMPLIFIED LINEAR GRAPH MODEL OF A TRANSLATIONAL MASS

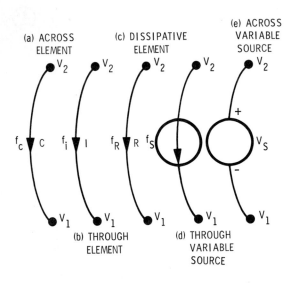

(a) ACROSS ELEMENT

(b) THROUGH ELEMENT

(c) DISSIPATIVE ELEMENT

(d) THROUGH VARIABLE SOURCE

(e) ACROSS VARIABLE SOURCE

Figure 9-80 Generalized schematic diagrams (linear graphs).

equations in one place for easy reference. The table also shows that the linear graph representation of the ideal elements is simpler to use than the original conceptual model.

In order to illustrate how these linear graphs can be used to replace the conceptual models, we will formulate several example problems in terms of the linear graphs. Figure 9-81b is a dynamic model using linear graph elements. The arrow on the mass symbol is usually shown pointing down since the velocity of the mass, v, is taken to be positive with respect to $v_g = 0$. The direction of the arrows on the spring is arbitrary and merely defines whether we are calling a tension or compression force as positive. Figure 9-82 is a linear grasp version of the automobile suspension (Figure 9-23). Note that the arrows on the masses point toward the reference velocity,

Mechanical example: (Figure 9-19)

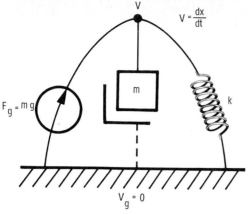

(a) DYNAMIC MODEL USING CONCEPTUAL ELEMENTS

(b) DYNAMIC MODEL USING LINEAR GRAPH ELEMENTS

Figure 9-81

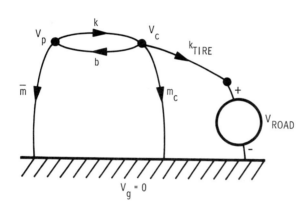

Figure 9-82 Linear graph model of an automobile suspension.

Electrical example: (Figure 9-47)

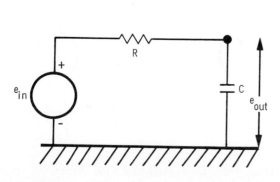

Figure 9-83a Dynamic model of a low-pass filter.

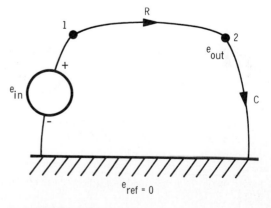

Figure 9-83b Linear graph of a low-pass filter.

Fluid example: (Figure 9-61)

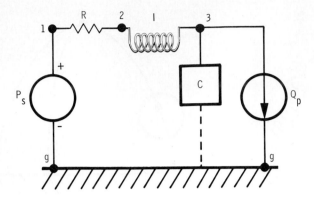

Figure 9-84a   Dynamic model of a water supply system.

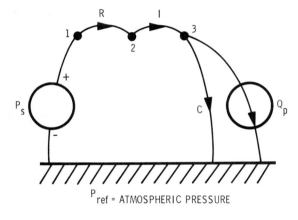

Figure 9-84b   Linear graph of a water supply system.

P_{ref} = ATMOSPHERIC PRESSURE

Thermal example: (Figure 9-73)

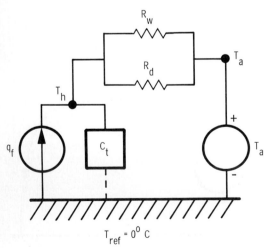

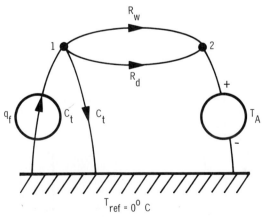

Figure 9-85a   Dynamic model of a home heating system.

Figure 9-85b   Linear graph of a home heating system.

$v_G = 0$. The directions of the arrows on the two springs and the damper are arbitrary as long as the correct elemental equation is used. For example, the direction of the arrow on the tire spring requires that

$$\frac{df_{tire}}{dt} = (k_{tire})(v_c - v_{road})$$

Figures 9-82 to 9-85 show how the use of linear graphs can simplify the formulation of conceptual models. The real power of the linear graph formulation, however, is that it also simplifies *equation formulation*. Up to this point we have identified a number of ideal elements and have shown how they can be connected together to model physical systems. If we are to use these models to *predict how the system will behave*, we must be able to *formulate the system equations*.

Problems

9-70. Sketch the linear graph of Problem 9-12.
9-71. Sketch the linear graph of Problem 9-13.
9-72. Sketch the linear graph of Problem 9-15.
9-73. Sketch the linear graph of Problem 9-16.
9-74. Sketch the linear graph of Problem 9-17.
9-75. Sketch the linear graph of Problem 9-30.
9-76. Sketch the linear graph of Problem 9-31.
9-77. Sketch the linear graph of Problem 9-50.
9-78. Sketch the linear graph of Problem 9-51.
9-79. Sketch the linear graph of Problem 9-52.
9-80. Sketch the linear graph of Problem 9-65.
9-81. Sketch the linear graph of Problem 9-66.

Equation formulation

Figures 9-81 to 9-85 represent *linear graph* models of practical physical systems and are composed of *ideal elements* that are interconnected. From the previous development and Table 9-1 we know the *elemental equations* for these ideal elements. For instance, in Figure 9-82 the elemental equation for the ideal mass tells us how the force acting through the case mass is related to the velocity across it; that is, $f_{m_c} = m_c(dv_{m_c}/dt)$. However, a more interesting problem concerning the typical shock absorber (shown in Figure 9-20) is to determine what the vertical velocity (v_p) of the passenger compartment will be when the tire is subjected to a vertical input, v_{road}. Figure 9-82 shows these two variables, *input* (v_{road}) and *output* (v_p). Likewise in Figure 9-83, we might like to be able to determine what the voltage acorss the capacitor, e_{out}, is when an arbitrary input voltage, e_{in}, is applied.

In order to determine these relationships we need to develop two additional concepts, the concepts of *compatibility* and *continuity*.

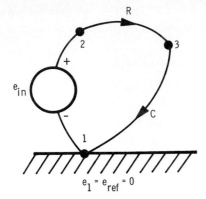

Figure 9-86   Linear graph of a low-pass filter.

Compatibility

The concept of compatibility is rather simple but very necessary and useful. Simply stated, the principle of compatibility states that the across variable at a specified point equals itself. For example, revising Figure 9-83 as Figure 9-86,[19] the voltage of point 2 with respect to point 1 is merely the difference in the voltages; that is,

$$e_{21} = e_2 - e_1 \tag{9-40}$$

where the subscript notation e_{21} means 2 with respect to 1. Similarly,

$$e_{23} = e_2 - e_3 \tag{9-41}$$

and

$$e_{31} = e_3 - e_1 \tag{9-42}$$

We know from our element definitions that[20]

$$e_{21} = e_{\text{in}}$$
$$e_{23} = e_R$$
$$e_{31} = e_C$$

If we substitute the expression for e_1 in equation 9-40 into equation 9-42 and the expression for e_2 in equation 9-41 into equation 9-42, we find that

$$-e_{\text{in}} + e_R + e_C = 0 \tag{9-43}$$

This expression, although very useful, simply says that $e_1 = e_1$. We could obtain the same expression by starting at any point on the *loop*[21] of Figure 9-86 and sum the across variables until we return to where we originally started. The net change in the across variable will be zero. For example, if we start at *node*[22] 1 and proceed around the loop in a clockwise fashion, summing voltages until we return to node 1, we have

$$-e_{\text{in}} + e_R + e_C = 0 \tag{9-44}$$

---

[19] Note here that a single "node" has been formed from the base of the linear graph since there is no element between them. Also the nodes have been renumbered.

[20] Remember that the direction of the arrow implies an across variable *decrease* in that direction. Also the notation e_R means the voltage across the resistor.

[21] A loop is defined as any closed path of a linear graph.

[22] A node is defined as any point where two or more elements are connected.

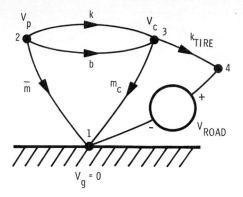

Figure 9-87 Linear graph of an automobile suspension.

Notice that, when we pass through an element in the direction of the arrow, there is a *positive* across variable drop, and, similarly, if we pass through an element with an arrow in the opposite direction, there is a negative across variable drop. Summarizing, we see that the *compatibility* relations are equations which relate the *across* variables around any loop of a linear graph. In electrical systems the compatibility relation is called Kirchhoff's voltage law.

In order to obtain some practice with compatibility, the four examples discussed previously will be treated.

Example: automobile suspension   Figure 9-87 is a repeat of Figure 9-82 with the four nodes numbered. In this problem we have four nodes and a possibility of six simple[23] loops. Figure 9-88 illustrates these simple loops. We can write compatibility equations for all the loops shown in Figure 9-88. Doing this, we obtain

loop (a)	$-v_p + v_k + v_c = 0$
loop (b)	$-v_p + v_b + v_c = 0$
loop (c)	$v_k - v_b = 0$
loop (d)	$-v_p + v_k + v_{k_\text{tire}} + v_\text{road} = 0$
loop (e)	$-v_p + v_b + v_{k_\text{tire}} + v_\text{road} = 0$
loop (f)	$-v_c + v_{k_\text{tire}} + v_\text{road} = 0$

$$(9\text{-}45)$$

By looking at equations 9-45 we see that only three of the six equations are *independent*. For example, by subtracting the second equation from the first we obtain the third, and by subtracting the fifth from the fourth we also obtain the third,

Example: water supply system   Figure 9-84b is the linear graph of the water supply system. There are three simple loops for this system, and writing a compatibility equation for each one yields

$$-P_S + P_R + P_I + P_C = 0$$
$$-P_S + P_R + P_I + P_P = 0$$
$$-P_C + P_P = 0$$

$$(9\text{-}46)$$

The term P_P represents the pressure drop across the pump. Note that, although the flow through the pump is specified, the pressure across it is not. Thus unless we are

[23]A simple loop is a closed path that does not pass through the same node more than once.

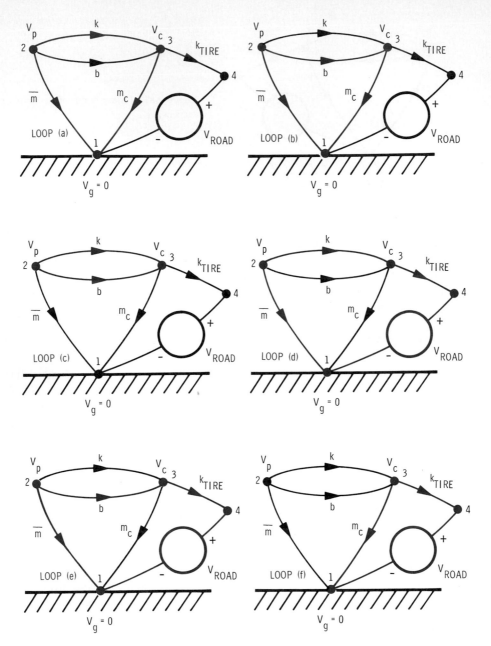

Figure 9-88 Simple loops.

interested in this quantity the last two of equations 8-46 provide no new information and in general only the first equation need be written. Generalizing this we can say that *it is unnecessary to write the compatibility equation for loops that contain a through variable source.* The first equation relates the constant pressure of the reservoir (P_S), the pressure drop across the resistance of the pipe (P_R), the pressure drop across the inertance of the pipe (P_I), and the pressure drop across the storage tank (P_C). Note that the pipe has been modeled as a series combination of a resistance and a fluid inertance; thus the pressure drop across the pipe is the sum of the two pressure drops ($P_R + P_I$).

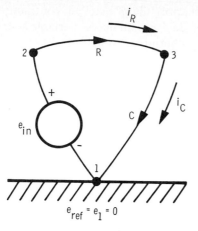

Figure 9-89 Linear graph of an electrical low-pass filter.

Example: home heating system Figure 9-85b is the linear graph of a home heating system. The necessary compatibility relations[24] are

$$-T_{C_t} + T_{R_d} + T_A = 0 \tag{9-47}$$
$$T_{R_w} - T_{R_d} = 0 \tag{9-48}$$

The first equation states that the temperature of the house ($T_{C_t} = T_h$) is equal to the outside temperature (T_A) plus the temperature drop across the door (T_{R_d}). The second equation simply states that the temperature drop across the window (T_{R_w}) is equal to the temperature drop across the door (T_{R_d}).

Continuity

We have developed the concept of *compatibility*, which relates the *across variables* of the system. We now develop the concept of *continuity*, which relates the *through variables* of the system. Figure 9-89 shows the linear graph of a low pass filter. Notice at *node 3* of the graph that the resistor current, i_R, is passing *into* the node, while the capacitor current, i_C, is passing *out* of the node. Since we know that electric charge (and consequently the current) cannot be created or destroyed, we find that $i_R = i_C$.[25] The concept of continuity is a statement of *conservation;* it simply says that at every *node* in a linear graph the sum of the *through* variables *into* the node must equal the sum of the *through* variables *out* of the node. For the electrical system, continuity is known as Kirchhoff's current law.

We can apply the continuity principle at every node in the linear graph; however, just as we did not obtain any useful information by applying compatibility around a loop that contained a through variable source, in like manner we do not obtain any useful information by applying the principle of continuity to a node that has an across variable source connected to it. Thus, in Figure 9-89, if we applied continuity to node 2, we would find that the current through the voltage source was equal to the current through the resistor. Although this is certainly true, it is the voltage across the source that is known, not the current through it.

[24]The compatibility equations for the loops containing a through variable source have been left out.
[25]The sign convention of the linear graph is that the through variable flows in the direction of the arrow.

Example: automobile suspension We will apply the principle of continuity to nodes 2 and 3 of Figure 9-87. Note that since node 4 has an across variable source (v_{road}) connected to it there is no need to apply continuity to it.

$$\boxed{\text{node 2}} \qquad -f_{\overline{m}} - f_b - f_k = 0 \qquad\qquad (9\text{-}49)$$

$$\boxed{\text{node 3}} \qquad f_k + f_b - f_{m_c} - f_{k_{\text{tire}}} = 0 \qquad (9\text{-}50)$$

The signs are all negative in equation 9-49 since all the forces at node 2 in Figure 9-87 are leaving the node.

Example: water supply system Applying continuity to nodes 2 and 3 of Figure 9-84b yields

$$\boxed{\text{node 2}} \qquad Q_R - Q_I = 0 \qquad\qquad (9\text{-}51)$$

$$\boxed{\text{node 3}} \qquad Q_I - Q_C - Q_P = 0 \qquad (9\text{-}52)$$

Equation 9-52 expresses the fact that the net flow into the storage tank (Q_C) is equal to the flow from the pipe (Q_I) minus the flow through the constant displacement pump (Q_P).

Example: home heating system Applying continuity to the only node of Figure 9-85b *that does not contain an across variable* source connection (node 1) yields

$$\boxed{\text{node 1}} \qquad q_f - q_{R_w} - q_{R_d} - q_{C_t} = 0 \qquad (9\text{-}53)$$

This equation states that the net heat flow rate into the house (q_{C_t}) is equal to the heat flow rate of the furnace (q_F) minus the heat flow rate through the window (q_{R_w}) minus the heat flow rate through the door (q_{R_d}).

Equation formulation

We have now developed all the concepts needed to be able to formulate the equations that describes a system's *dynamic* behavior. We now develop a unified way to express them. The three required steps are as follows:

1. Write all the *elemental equations*.
2. Write the necessary *compatibility equations*.
3. Write the necessary *continuity equations*.

Example: automobile suspension (Figure 9-87) We have already developed all the pieces to this problem.

1. *Elemental equations*

$$\frac{dv_c}{dt} = \frac{1}{m_c} f_{m_c} \qquad\qquad (9\text{-}54)$$

$$\frac{dv_p}{dt} = \frac{1}{\overline{m}} f_{\overline{m}} \qquad\qquad (9\text{-}55)$$

$$\frac{df_k}{dt} = kv_k \qquad\qquad (9\text{-}56)$$

$$f_b = bv_b \qquad\qquad (9\text{-}57)$$

$$\frac{df_{k_{\text{tire}}}}{dt} = k_{\text{tire}} v_{k_{\text{tire}}} \tag{9-58}$$

2. *Compatibility equations*

loop (a)	$-v_p + v_k + v_c = 0$	(9-59)
loop (c)	$v_k - v_b = 0$	(9-60)
loop (d)	$-v_p + v_k + v_{k_{\text{tire}}} + v_{\text{road}} = 0$	(9-61)
loop (f)	$-v_c + v_{k_{\text{tire}}} + v_{\text{road}} = 0$	(9-62)

3. *Continuity equations*

node 2	$-f_{\bar{m}} - f_b - f_k = 0$	(9-63)
node 3	$f_k + f_b - f_{m_c} - f_{k_{\text{tire}}} = 0$	(9-64)

Equations 9-54 through 9-64 represent a complete set of *differential equations,* that is, a set of equations in terms of the system variables and their *derivatives.* These equations can be expressed more compactly by manipulating the equations so that they contain only the variables whose derivative appears in the elemental equations and the input sources. From equations 9-54 through 9-58 we see that the variables whose derivatives appear are v_c, v_p, f_k, and $f_{k_{\text{tire}}}$. Thus we can use all the other equations to eliminate v_b, v_k, $v_{k_{\text{tire}}}$, $f_{\bar{m}}$, f_{m_c}, and f_b. By making the appropriate *algebraic* substitutions, we have

$$\frac{dv_c}{dt} = \frac{1}{m_c}[f_k - f_{k_{\text{tire}}} + b(v_p - v_c)] \tag{9-65}$$

$$\frac{dv_p}{dt} = \frac{1}{\bar{m}}[b(v_c - v_p) - f_k] \tag{9-66}$$

$$\frac{df_k}{dt} = k(v_p - v_c) \tag{9-67}$$

$$\frac{df_{k_{\text{tire}}}}{dt} = k_{\text{tire}}(v_c - v_{\text{road}}) \tag{9-68}$$

Equations 9-65 through 9-68 are the final mathematical equations that can be used to predict how the automobile suspension will *behave dynamically* when subjected to an arbitrary road input, v_{road}. The *solution* of these *equations* to a specified forcing function (v_{road}) is beyond the scope of this chapter. However, differential equations of this form can sometimes be solved *analytically,* and they can always be solved using a digital or analog computer.

Example: electrical low pass filter (Figure 9-83b)

1. *Elemental equations*

$$\frac{de_c}{dt} = \frac{1}{C}i_c \tag{9-69}$$

$$e_r = Ri_R \tag{9-70}$$

2. *Compatibility*

$$-e_{\text{in}} + e_R + e_C = 0 \tag{9-71}$$

3. *Continuity*

$$\boxed{\text{node 2}} \qquad i_R - i_c = 0 \qquad\qquad (9\text{-}72)$$

As in the case of the shock-absorber equations, we can simplify equations 9-69 and 9-70 by retaining only the differentiated variable (e_C) and the source variable (e_{in}). Applying algebra to reduce the number of equations yields

$$\frac{de_c}{dt} = \frac{1}{RC}(e_{in} - e_C) \qquad\qquad (9\text{-}73)$$

Equation 9-73 is the fundamental equation that describes the low-pass filter behavior. Thus, once $e_{in}(t)$ is specified,[26] we can solve for $e_C(t)$.

Example: water supply system (Figure 9-84)

1. *Elemental equations*

$$\frac{dQ_I}{dt} = \frac{1}{I}P_I \qquad\qquad (9\text{-}74)$$

$$\frac{dP_C}{dt} = \frac{1}{C}Q_C \qquad\qquad (9\text{-}75)$$

$$P_R = RQ_R \qquad\qquad (9\text{-}76)$$

2. *Compatibility*

$$-P_S + P_R + P_I + P_C = 0 \qquad\qquad (9\text{-}77)$$

3. *Continuity*

$$\boxed{\text{node 2}} \qquad Q_R - Q_I = 0 \qquad\qquad (9\text{-}78)$$
$$\boxed{\text{node 3}} \qquad Q_I - Q_C - Q_P = 0 \qquad\qquad (9\text{-}79)$$

Expressing these equations in terms of the differentiated variables (Q_I, P_C) and the source variables (P_S, Q_P) yields

$$\frac{dQ_I}{dt} = \frac{1}{I}(P_S - RQ_I - P_C) \qquad\qquad (9\text{-}80)$$

$$\frac{dP_C}{dt} = \frac{1}{C}(Q_I - Q_P) \qquad\qquad (9\text{-}81)$$

Equations 9-80 and 9-81 are the fundamental equations that define how the water supply system will respond to a specified reservoir pressure (P_S) and a specified pump flow (Q_P).

Example: home heating system (Figure 9-85b)

1. *Elemental equations*

$$\frac{dT_{C_t}}{dt} = \frac{1}{C_t}q_{C_t} \qquad\qquad (9\text{-}82)$$

[26] The notation $e_{in}(t)$ means that e_{in} is a function of time.

$$T_{R_w} = R_w q_{R_w} \qquad (9\text{-}83)$$

$$T_{R_d} = R_d q_{R_d} \qquad (9\text{-}84)$$

2. *Compatibility*

$$-T_{C_t} + T_{R_d} + T_A = 0 \qquad (9\text{-}85)$$

$$T_{R_w} - T_{R_d} = 0 \qquad (9\text{-}86)$$

3. *Continuity*

$$\boxed{\text{node } 1} \qquad q_f - q_{R_w} - q_{R_d} - q_{C_t} = 0 \qquad (9\text{-}87)$$

Again, we will express these equations in terms of the differentiated variable (T_C) and the source variables (q_f, T_A). The appropriate algebraic manipulations yield

$$\frac{dT_{C_t}}{dt} = \frac{1}{C_t}\left(\frac{1}{R_w} + \frac{1}{R_d}\right)(T_A - T_{C_t}) + \frac{q_f}{C_t} \qquad (9\text{-}88)$$

Equation 9-88 is the fundamental equation for the home heating system. It describes how the temperature of the house $(T_h = T_{C_t})$ will change when subject to a varying atmospheric temperature (T_A) and an adjustable furnace heat flow rate (q_f).

Example: qualitative model for the human heart   Figure 9-90 is a schematic representation of the heart and associated circulatory system. Suppose it is desired to investigate in a qualitative way the effect of smoking, hardening of the arteries, and

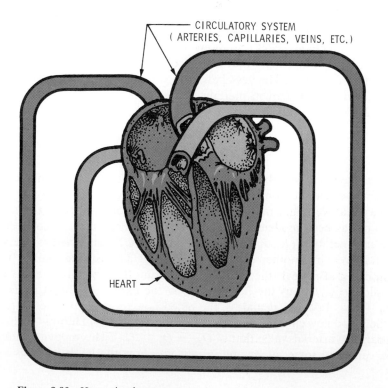

Figure 9-90   Heart-circulatory system.

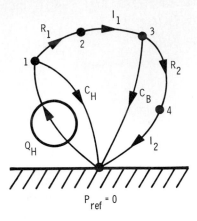

Figure 9-91 Linear graph representation of an idealized model of a heart artery system.

cholesterol on the blood pressure. We will need to develop an engineering model that is simple, and yet it will need to incorporate as many of the important characteristics of the circulatory system as possible. We know that primarily the heart acts as a pump; therefore, we will need to include a *volume flow rate source, Q_H*. The heart also is elastic and can store energy as a *fluid capacitor, C_H*. The artery and capillary system can be modeled as small pipes that offer *resistance* to flow and also *inertance* (store energy due to the flow through it). We also know that the blood vessels are elastic and therefore our model should include some *capacitance*.

The modeling of the artery and capillary system requires careful evaluation. A long thin elastic tube appears to possess all three ideal element properties. It can store energy by virtue of a pressure difference; in this case it is the elastic nature of the vessel walls that expand under pressure (see Figure 9-54). Clearly the capacitance of the blood vessel is proportional to its elasticity since the capacitance decreases as the artery stiffens. The blood vessel also offers a resistance to flow, which for long thin tubes is given by equation 9-31; that is, $R = G_1/d^4$, where G_1 is a constant[27] and d is the diameter of the tube. The vessel also can store energy by virtue of the volume flow rate through it, that is, by its inertance. The inertance of a long tube is given by equation 9-28:

$$I = \frac{\rho l}{A} = \frac{\rho l}{\pi d^2/4} = \frac{G_2}{d^2}$$

Now that we have assembled a set of ideal fluid elements that characterize the heart system, we need to connect them together properly. Figure 9-91 shows one possible way that we might piece together the ideal elements. Note that the blood vessel system has been modeled as five lumped elements: a series connection of a resistance and inertance (R_1, I_1), and a capacitance (C_B) connected to the reference pressure, followed by another series connection of a resistance and an inertance (R_2, I_2). Looking at node 1, the blood flow entering the arterial system is equal to the heart flow (Q_H) minus the flow stored by the expanding heart walls; that is, $Q_{R_1} = Q_H - Q_{C_H}$. Looking at node 3, the blood flow returning to the heart (Q_{R_2}) is equal to the blood flow entering the arterial system ($Q_{R_1} = Q_{I_1}$) minus the flow stored by the expanding arterial walls; that is, $Q_{R_2} = Q_{I_1} - Q_{C_B}$. This model, although very simplistic, can yield some interesting insight into how the heart behaves.

---

[27]For a particular situation the value $128\mu l/\pi$ can be represented by a constant, for example G_1.

To obtain the complete set of equations for the heart system the elemental, compatibility, and continuity equations are written as follows:

Elemental Equations	Compatibility	Continuity
$\dfrac{dP_{C_H}}{dt} = \dfrac{1}{C_H} Q_{C_H}$	$-P_{C_H} + P_{R_1} + P_{I_1} + P_{C_B} = 0$	$Q_H = Q_{C_H} + Q_{R_1}$
$\dfrac{dQ_{I_1}}{dt} = \dfrac{1}{I_1} P_{I_1}$		$Q_{R_1} = Q_{I_1}$
$\dfrac{dP_{C_B}}{dt} = \dfrac{1}{C_B}(Q_{C_B})$	$-P_{C_B} + P_{R_2} + P_{I_2} = 0$	$Q_{I_1} = Q_{C_B} = Q_{R_2}$
$\dfrac{dQ_{I_2}}{dt} = \dfrac{1}{I_2} P_{I_2}$		$Q_{R_2} = Q_{I_2}$
$P_{R_1} = R_1 Q_{R_1}$		
$P_{R_2} = R_2 Q_{R_2}$		

Again by expressing these equations in terms of the variables whose derivative appears in the elemental equations, we obtain

$$\frac{dP_{C_H}}{dt} = \frac{1}{C_H}(Q_H - Q_{I_1}) \tag{9-89}$$

$$\frac{dQ_{I_1}}{dt} = \frac{1}{I_1}(P_{C_H} - P_{C_B} - R_1 Q_{I_1}) \tag{9-90}$$

$$\frac{dP_{C_B}}{dt} = \frac{1}{C_B}(Q_{I_1} - Q_{I_2}) \tag{9-91}$$

$$\frac{dQ_{I_2}}{dt} = \frac{1}{I_2}(P_{C_B} - R_2 Q_{I_2}) \tag{9-92}$$

Applying compatibility around the heart yields $P_{C_H} = P_H$; thus equations 9-89 to 9-92 determine the system output (P_H) as a function of Q_H and the blood vessel diameter, d.

We will make the assumptions that smoking increases the heart flow rate, Q_H, that hardening of the arteries decreases the blood vessel capacitance, C_A, and that cholesterol decreases the blood vessel diameter, d.

A careful analysis of equations 9-89 to 9-92 would indicate to us that increasing Q_H, decreasing C_H, and decreasing d would *increase* the heart pressure. A definitive solution of equations 9-89 to 9-92 is beyond the scope of this chapter. However, it is important that we recognize that a complex system, like the heart, can be modeled by the principles described. It should be emphasized that the heart model developed above is a *very simple* and idealized model, and it should be used only to gain a qualitative insight into how the heart model is affected by certain stipulated parameters.

An example of an analytical solution

In general the differential equations that are formulated for a given system need to be solved by the use of a computer. However, if the model is simple enough the solution often can be found by analytical means. For example, the fundamental

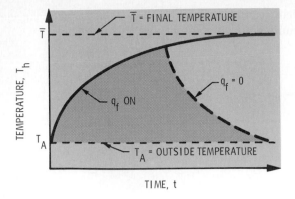

Figure 9-92   Temperature-time history.

equation for the home heating system (equation 9-88) is a *first-order* differential equation. If we assume that the atmospheric temperature and the furnace heat flow rate are constant, then the complete solution is given by

$$T_h = R_{eq}q_f(1 - e^{-t/R_{eq}C_t}) + T_A$$

where $R_{eq} = R_w R_d/(R_w + R_d)$, and is defined as the equivalent resistance of the combined parallel window and door resistances. This solution assumes that the initial temperature of the house is the outside atmospheric temperature (T_A) before the furnace is turned on (q_f). A typical *time history* of the house temperature is shown in Figure 9-92.

The final temperature ($\overline{T}$) that the house will eventually reach can be found by setting $dT_H/dt = 0$ in equation 9-88, or equivalently by setting $t = \infty$ in the temperature solution. This yields

$$\overline{T} = T_A + R_{eq}q_f$$

Therefore, the final temperature is a function of the atmospheric temperature, the equivalent resistance, and the furnace heat flow rate. The speed with which this final temperature is achieved is dependent on the equivalent resistance and the thermal capacitance, C_t.

If at any time the furnace is turned off ($q_f = 0$), the temperature will begin to decrease, as shown in Figure 9-92.

Suppose it is desired to determine what the effect of decreasing the equivalent resistance (e.g., opening the window slightly) would be on the house temperature. All

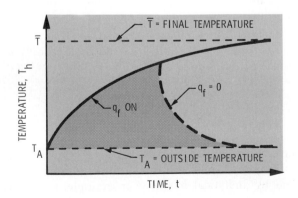

Figure 9-93   Temperature-time history for reduced R_{eq}.

we have to do is to decrease the term R_{eq} in the complete time solution. Figure 9-93 shows a sketch of the resulting temperature time history.

Note that the *final temperature* achieved is lower than before and that the *rate* of increase and decrease of temperature is greater than before. Thus, if for a given furnace we want a high maximum temperature capability and also want the house to dissipate its heat slowly, it is important to have as high an R_{eq} as possible.

Problems

9-82. The fundamental equation for the electrical low pass filter is given by equation 12-73. Suppose that e_{in} = constant = 6 volts. What will be the *steady-state* value of e_c?

9-83. At $t = 0$ seconds, the voltage across the capacitor in Figure 9-83b is equal to zero volts. Assuming that e_{in} = 6 volts, sketch the *time response of the system*. (*Hint:* notice the similarity between this problem and the thermal example.)

9-84. Assuming the same conditions as in Problem 9-83, sketch how the time response of $e_c(t)$ would change if (*a*) RC is increased; (*b*) RC is decreased.

9-85. Using equation 9-88, predict what effect on the final temperature of the house the following parameter changes would have (assuming T_A and q_f are constant): (*a*) an increase in R_d; (*b*) decrease in C_t.

9-86. Using equation 9-88, predict what effect on the *rate of change of house temperature* (dT_h/dt) the following parameter changes would have: (*a*) increase in R_w; (*b*) increase in C_t.

9-87. The fundamental equations for the water supply system are 9-80 and 9-81. Find the steady-state values of Q_1 and P_C assuming the values of C, Q_P, P_S, R, and I are known. (*Hint:* Set rate terms equal to zero.)

9-88. What will be the steady-state value of the storage tank pressure (P_C) if the pump flow (Q_P) is equal to zero?

9-89. What would be the effect of increasing the storage tank capacitance (c) on the *rate of change of pressure* (dP_C/dt)?

9-90. What would be the effect on the steady-state tank pressure (P_C) if the pipe diameter were increased?

In Problems 9-91 through 9-102, (1) sketch the linear graph, (2) write the elemental equations, (3) write the compatibility equations, (4) write the continuity equations, and (5) formulate the *system equations* in terms of only the variables that appear as derivatives in the elemental equations and the input sources.

9-91. Use the system shown in Problem 9-12.
9-92. Use the system shown in Problem 9-13.
9-93. Use the system shown in Problem 9-15.
9-94. Use the system shown in Problem 9-16.
9-95. Use the system shown in Problem 9-17.
9-96. Use the system shown in Problem 9-30.
9-97. Use the system shown in Problem 9-31.
9-98. Use the system shown in Problem 9-50.
9-99. Use the system shown in Problem 9-51.
9-100. Use the system shown in Problem 9-52.
9-101. Use the system shown in Problem 9-65.
9-102. Use the system shown in Problem 9-66.

In the following four problems, (1) formulate a simple dynamic model, (2) sketch the linear graph, (3) find the fundamental equations, and (4) analyze the effect of the basic elements in your model.

9-103. A new vehicle is to be tested for its crashworthiness. Important elements to include in your model are the front bumper (stiffness k_b and damping b_b), the driver (mass m_d), the restraining harness (stiffness k_H), and the vehicle mass (m_v).

9-104. A drainage system has been proposed for a city. The primary components of the system are as follows. A large storage tank (C_1) is located at one part of the city. After a large rainfall the water (Q_1) is channeled into this storage tank. A long underground pipe (I, R) connects this storage tank to a second storage tank (C_2) at the outskirts of the city. A pumping station (Q_{out}) is connected to this second storage tank and pumps the water out of the tank into an open area.

9-105. It is desired to build an electrical circuit that can simulate the behavior of the new bumper described in Problem 9-103. Illustrate how this could be done.

9-106. A home heating system contains the following elements: a furnace (q_f), a radiator (thermal resistance R_r), the room (C_t, R) and the external temperature (T_A).

10

Designs in nature

What evidence would suggest that the inferior cockroach might outlast man, the most advanced creature that the world has known? Isn't man the most intelligent of all creations, and doesn't he adapt himself readily to changing geographic and climatic conditions, and to unusual and unique environments that are entirely foreign to his natural habitat? Aren't his recent journeys to the moon an excellent example of this ability? Yes, these are all true statements, and there are even stronger statements that could be made in behalf of man. For example, only man has been given intellectual faculties and a spiritual nature in addition to physical qualities. He alone possesses the ability to reason and he has lived a long, long time. In this regard, many scientists believe that man may have inhabited the earth for almost 2 million years.

In the works of Nature *purpose,* not accident, is the main thing.
—Aristotle *Nicomachean Ethics*

If no way be better than another—that, you may be sure, is Nature's way.
—Leibniz *Théodicéc,* 1857

The chessboard is the world, the pieces are the phenomena of the universe, the rules of the game are what we call the laws of Nature. The player on the other side is hidden from us, we know that his play is always fair, just, and patient. But also we know, to our cost, that he never overlooks a mistake, or makes the smallest allowance for ignorance.
—Thomas Henry Huxley *A Liberal Education,* 1825–1895

Illustration 10-1

And when the span of man has run its course
Sitting upon the ruins of his civilization will be a cockroach
Calmly preening himself
In the rays of the setting sun.

However, the cockroach has a much better "track record" for staying power. He has been around some 280 million years, or about 140 times as long as man. During man's *brief* tenure on earth the cockroach has competed successfully with him for food, lodging, and other necessities of life. Today, unlike numerous other animals and insects, his extinction does not appear to be imminent. Rather he seems to be gaining strength with each passing generation. Upon examination we find that the roach has been designed as a finely tuned instrument capable of adjusting to a tremendous range of conditions. He has a full complement of structures, organs, and behavior patterns that have been tested and honed to a fine edge by the evolutionary forces of nature. Although his physical size is small, his internal mechanisms are considerably more complex than the most sophisticated products of man's modern technology. Admittedly, he lacks intelligence and reasoning power, but he has never been observed to engage wilfully in activities that would guarantee his eventual annihilation. Yes, it *is* possible that man might learn many things from observing the actions of the lowly cockroach.

It is now believed that the earth is at least 3 billion years old. It is filled with many types of living things—animals and insects and plant life of countless varieties. Like the cockroach, each occupies a special place and each acts in a predictable behavior pattern in accordance with a long experience in meeting the requirements of nature under varying environmental conditions. To survive, each animal, insect, and plant has been forced to contrive actions, solutions, and *designs* that provide protective housing, that enhance the probability of sensing and capturing food, and that locate either prey or predator. Many of these designs are ingenious, and a large number are astonishingly beautiful. Many are simple designs. Others are quite complex, yet very efficient.

It is unfortunate that most often man has acted only as a casual observer of these ingenious *designs of nature*. More rarely, he has been a user of the products of nature's design (such as diatomic earth, spider-web threads, and bamboo). Least frequently, he has been an imitator of nature and been able to make use of the unique solutions that nature has made available to us (such as flying machines, honeycomb structures, and poison arrows). In fact, in the majority of cases man often has discovered the existence of some unique design in nature only after he himself had invented a similar device and learned to recognize its qualities.

The engineering student should be especially observant of the behavior of animals, insects, and plants, to make an effort to discover how these "living designs of nature" have solved their problems, and to make use of this information in designing for human preservation and satisfaction. Some examples of designs in nature that are being used by man are described below.

The Sonar of bats

Long before we had heard of *sonar*, bats were using it to navigate and locate their prey. Insectivorous bats are able to locate and to capture insects smaller than mosquitoes while flying in complete darkness at full speed. They do so by giving off a series of ultrasonic cries and locating objects by the distance and direction of the sources of echoes.

The sounds emitted by bats are actually very loud with intensities as high as 113 decibels. We are unable to hear them because they are out of the range of our sound sensitivities. The frequency modulation of the pulses ranges from 20,000 to 100,000 cycles per second. Man cannot hear sounds above 20,000 cycles per second. The wavelength of the pulses varies from 6 to 12 millimeters, which is about the same size as the insects that make up the normal prey of the bats.

While hunting, the bat sends 10 to 20 pulses per second until a returning echo tells him that an object is in his flight path. He then increases the pulse rate to as high as 200 per second. This improves his discrimination and enables him to find the prey more accurately.

A great deal of interest has been generated in the bat's "sonar" system because of two of its characteristics that are as yet unexplained. Some bats are able to locate fish beneath the surface of the water, and it is not understood how this is done. Also, attempts to "jam" the bat's sonar system with loud sound intensities have not been successful. In some way the bat is able to detect its own signals through a cacophony of sound. Military men have been looking for jam-proof systems and consequently are very interested in learning how the bat is able to accomplish this feat. Another use has been made of the principle in the designing of a sonar system to aid blind people to "see."

It is interesting to note that some night-flying moths which are the normal prey of bats have hearing apparatus that can detect the ultrasonic cries of the bats. When they hear a bat approaching, they may drop to the ground, soar upward, dive steeply, or otherwise take evasive action. The moth can detect the bat at a greater

Illustration 10-2

The bat is the only flying mammal. For food it hunts insects at night, and it finds them by an intricate sonar system. It issues periodic beeps of constant amplitude (as high as 113 decibels) but varying frequency— 30,000–60,000 hertz—well beyond the range of human ears. The bat can tell from the echo reflected by its prey exactly how far away the prey is and where it is headed. This means that its ears must be extremely sensitive for receiving but unaffected by the loud sending signal. Careful study of the bat's ultrasonic sonar system has enabled engineers to design a device that allows blind people to "see."

distance than the bat can detect it. However, the bat is capable of greater speed in flight. It is interesting that in this predatory–prey relationship there is a precise balance between the ability of the bat to locate and react to a moth and the ability of the moth to detect and react to the bat.

The eye of the frog

The eyes of frogs are much less acute than those of man, and they are not movable in their sockets. However, they do have one capability that is very important to the frog's ability to obtain food. That is, they are able to detect movement and to differentiate between things flying toward him and those flying away from him. An engineering version of the retina of the frog is currently being tested to determine its application as a device to assist air-traffic controllers to prevent air collisions. The model contains more than 33,000 electrical components to accomplish this action.

Flight control of flies

Gyroscopes have long been used to stabilize the position of structures and vehicles, such as ships and planes. They are based upon the principle of a rapidly spinning wheel or disc that is mounted so that it is free to rotate about one or both of two axes perpendicular to each other and to the axis of spin—thereby offering opposition to any torque that would change the direction of the axis of spin.

Illustration 10-3
Frogs are practically blind, and their eyes are stationary in their sockets. However, they have other remarkable abilities. They receive, process, and relay information important to the frog and filter out everything else. For example, a juicy bug flying toward him is important, one flying away is not. The frog's eye signals his brain if

(a) An object is flying toward him,
(b) The object is "bug size,"
(c) The object is flying at "bug speed,"
(d) The object is within range.

Everything else is ignored. The frog's eye also makes other life and death decisions for him without bothering his feeble brain. A sudden shadow, for example, will trigger a danger signal causing the reflexive jumping mechanism to function.

Illustration 10-4
The multifaceted compound eye of the common housefly (enlarged over 100 times here) produces a mosaic made up of multiple images. In much the same way, engineers have designed new multielement phased-array radar sets that are freed from the foibles of mechanical failure and possess a near-instantaneous ability to "look" from one direction to another.

Flies have only one pair of functional wings. However, a pair of small knobbed structures known as *halteres* act as a second set of modified wings. These halteres vibrate up and down when the insect is in flight. As long as the insect is in level flight there are no abnormal strains at the point where the halteres are attached to the body. If the flying insect is inclined or tipped, however, the plane in which the halteres are vibrating is changed and stresses are set up in the attachment points to the body. The fly is then able to adjust his position to normal. In effect, the halteres act as "alternating gyroscopes."

The snake's hypodermic fang

In this day of antibiotics in medicine it is difficult for anyone to escape the sting of the hypodermic needle. Even when being subjected to the needle, few people take the opportunity to meditate on the fact that the hypodermic needle is an engineering copy of the fangs of rattlesnakes, which have served the snake very efficiently for millions of years. Rattlesnake fangs have exceedingly sharp points to break the skin and allow the entry of the fangs to the underlying tissues; the wedge-shaped design of the fangs spreads the tissues and allows the entry of the poison into the wound from an opening in the side of the fangs rather than from the tip. In this way, the possibility of the fangs becoming plugged by the tissue of the victim is minimized. The hypodermic needle has been designed with the same adaptive modification.

Jet propulsion in nature

Even jet propulsion, which is so widely utilized to propel aircraft and boats, has its antecedent in designs of nature. One animal that occurs in tremendous numbers in the oceans of the world is the *squid*. Squids are very important in the food chain in the oceans. In order to move, they employ the engineering principles employed by jet propulsion. Water is drawn into a cavity within the squid that is surrounded by

Illustration 10-5
Man has much to learn from the snake. Over the years efforts have been made to imitate the design of the snake's fangs in the manufacture of hypodermic needles and to copy the heat-seeking infrared sensor mechanism of his nostril pits in missile-defense systems. What may offer even more valuable use, however, would be an understanding of his odor-air quality tester—the combined action of this flicking, darting tongue and a small organ of smell located in the roof of the mouth.

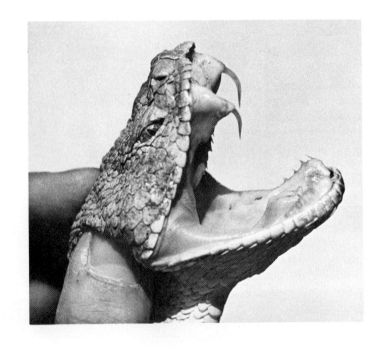

Illustration 10-6

Squids are remarkable creatures, and next to fish they are the most numerous of the advanced life forms of the ocean. Interesting features that we would do well to study are their jet-propulsion system, tentacles, beaks, dual optical systems (for seeing both in the darkened depths and on the surface), glowing light organs, adaptive camouflage, and black ink-like protective substance that serves to decoy and confuse attackers.

elastic muscular walls. When the muscles contract, the size of the cavity is decreased and the water is forced out through a funnel-shaped structure, thereby pushing the squid forward in accordance with Newton's third law of motion. By changing the direction in which the funnel is pointed, the squid is able to change its direction of movement.

Immature forms of dragonflies also employ jet propulsion for locomotion. These nymphs live in the water for 2 or 3 years before changing into the more familiar form of dragonflies. While in the water, the nymphs breathe by means of gills located in a cavity at the posterior end of the intestine. When the size of the cavity is increased, water enters. When the muscular walls of the cavity contract, water is forced out the posterior end of the animal and the nymph is forced forward. As a result of this intermittent action, the nymph moves in a sequence of short spurts.

Mechanisms and structures

Information gained from the study of animal movement has provided man with some of his most useful mechanical devices. Two of these are the hinge and ball-

Illustration 10-7
The dragonfly larva, or nymph, can walk or climb, and it swims efficiently by a jet-propulsion-like action. It captures its prey by means of a prehensile lower lip, which is folded under the head when not in use, but which can be extended suddenly to seize a prey by means of a pair of terminal hooks. Courtesy Carolina Biological Supply Company.

Illustration 10-8
Mechanical hinges and ball and socket connections are adaptations of similar designs within the human body.

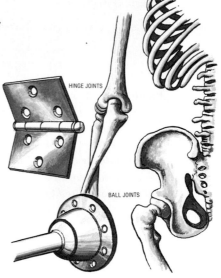

HINGE JOINTS

BALL JOINTS

and-socket joint. The hinge is represented by the wrist, elbow, and finger joints of man; the hip and shoulders are examples of ball-and-socket joints. Mechanical examples of hinges include door hinges and piano hinges. Mechanical copies of ball-and-socket joints include trailer hitches and ball joints on the steering systems of automobiles.

Perhaps the animal that has been most closely associated with civil-engineering exploits is the beaver. His prowess at building dams is almost legendary and certainly well deserved. Using logs, branches, stones, and mud, the beaver has built dams, complete with spillways, that are marvels of efficiency. Studies of beaver dams have revealed that they are bow shaped with the bow facing upstream to provide the greatest strength against the pressure of the water. Large man made dams also utilize this principle. Long before man constructed his first crude dam, beavers were build-

Illustration 10-9
With the tools at his disposal the beaver is a very capable design engineer.

ing coffer dams to divert water into canals to float logs to the desired dam construction site.

Living organisms have evolved some truly remarkable structures for specialized purposes. Among these are light-producing organs, electric organs, sonar systems, and other intricate guidance systems. Some of these natural systems are still not understood.

Cold light

Nearly everyone has seen fireflies and has marveled at their ability to produce light. However, most people are not aware that a great number of organisms have this ability, including bacteria, fungi, sponges, corals, insects, snails, crustaceans, clams, millipedes, and centipedes. The emission of light by organisms is called bioluminescence and is the result of an enzyme-catalyzed chemical reaction. The bioluminescent mechanism in different species is very similar, and it apparently evolved as a by-product of tissue metabolism. There has been a great deal of interest in the "cold light" produced by bioluminescence. Studies of its quantum efficiency have shown that firefly light production is almost 100 per cent efficient. Bioluminescence may be used as protection, as a lure, or as a mating signal. In some organisms its function is quite obscure. Colors produced by bioluminescence range over nearly the whole visible spectrum, with blue, blue-green, or white predominating. Green is the most frequent, but red and orange are relatively rare. There are three basic ways in which

Illustration 10-10

Fireflies, glow worms, or fire beetles produce their characteristic bioluminescent glow by means of a complex and delicate chemistry that is not yet fully understood. Luciferin (a biochemical substance) oxidizes in the presence of an enzyme called luciferase and this serves as the fuel for the firefly's lantern. Both substances are produced in the abdomen of the firefly, where they are fed oxygen from the atmosphere by means of microscopic breathing ducts. While the usual incandescent lamp wastes 98 per cent of the energy supplied as heat, the efficiency of the cold-light production plant of the firefly is amazingly high, almost 100 per cent.

bioluminescence may be produced: (1) by symbiotic bacteria housed in special organs, (2) by light-emitting cells, and (3) by production of luminous secretions.

Some squids and deep-sea fish have symbiotic bacteria that are kept in specialized sacs where they grow. The emission of light is controlled by specialized structures, such as pigmented curtains, lenses, or reflectors that concentrate or diminish the light.

Luminous secretions are common among marine invertebrates. Secretions may completely cover the body or may appear as a luminescent cloud or discrete points of light. This luminescence is usually intermittent, with the control in the nervous system of the animal. In most advanced forms, light-emitting cells produce light within the cell. The most familiar example is the firefly.

Electricity in nature

Electric charges are generated in all creatures with a nervous system. However, an external discharge of electricity has been noted only in certain groups of "electric" fish. This electricity is used for orientation, communication, and protection.

Electric organs are composed of electroplates, which are modified muscle cells. These plates are thin flattened structures whose two specialized surfaces are different from each other. There are large numbers of these cells, and those having similar surfaces are oriented in the same direction, horizontally or vertically. Each plate is housed in a compartment and embedded in a gelatinous material.

The electric eel has about 40 per cent of its body devoted to electroplates, ar-

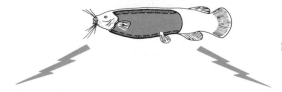

Figure 10-1 The electric catfish (Nile River).

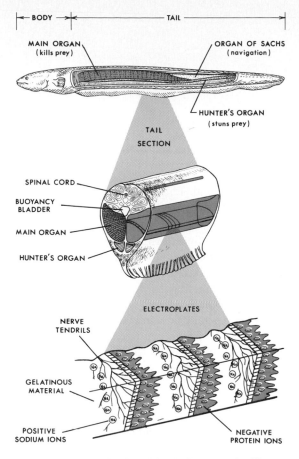

Figure 10-2 The electric eel has built up a radar-like system to detect its prey. Special elastic organs discharge pulses of electricity that create an oscillating dipole field capable of detecting anything entering the field and distorting the pattern. Conductors tend to concentrate the lines of electrical current flow, whereas insulators repel them. Big, healthy electric eels can produce a 0.003-second jolt that has been measured at 887 volts and 1 ampere. If we really understood how the eel's electrical system worked, we could design vastly improved electrical-power-generating systems.

ranged vertically in about 120 columns parallel to the spinal cord with each column containing 6,000 to 10,000 electroplates. The electroplates are arranged in series and are able to generate more than 500 volts. The high voltage is necessary to overcome the substantial electrical resistance of the freshwater in which the eel lives. On the other hand, the torpedo ray has electroplates stacked horizontally in about 2,000 columns with approximately 1,000 plates in each column. These columns are arranged in parallel, which enables the ray to develop a large amperage. This is most appropriate because the ray inhabits salt water, which offers a lower electrical resistance. The electric organs of both are used for stunning prey or repelling predators, and the electrical charge developed is large enough to be a considerable deterrent.

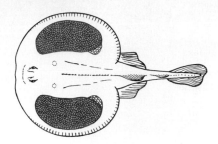

Figure 10-3 The torpedo ray (Pacific Ocean).

Some fish produce almost continuous streams of electric pulses, the frequencies ranging from 50 to 1,600 pulses per second. These pulses are used to orient the fish and to enable it to detect objects near it.

An African fish *Glymnarchus niloticus* is able to detect objects as small as glass rods 2 millimeters in diameter by means of an electrical mechanism. The mechanism consists of an electric generating organ in the tail of the fish that is negative relative to the fish's head. In the head region the skin is perforated with pores that lead into tubes filled with a jelly-like substance. The lines of the electric current emitted from the tail are focused on these pores at the head end of the fish. Any objects in the electric field will distort the lines of current flow and tell the fish that an object is present in the field. The time it takes the fish to average out the information from 7 or 8 discharges of the electric organ has been found to be 25 milliseconds. How these sensory cells are able to accomplish this is still unknown.

Unusual mechanisms

Propulsion is very much a part of our modern world. A number of examples of the use of rockets and jet engines quickly come to mind. However, all man-made propulsion devices now in use have some problems, such as pollution, excessive consumption of fuel, etc. For the movement of an animal or the propulsion of a weapon of prey or defense, a number of methods are known in nature that have none of these drawbacks and that, if completely understood, might have some application to use by man. Studies are now being carried out to investigate these methods.

Studies of how fish swim have indicated great similarities between the tail fin of fish and variable propeller blades used on ships and submarines. Fish, of course, have to use a system of levers to produce a propelling effect. This system of levers consists of the vertebral column together with the tail fin. In fish the tail fin moves back and forth laterally; in whales and dolphins the tail fin moves up and down. Regardless of which way it moves, the speed of the tail fin varies during one cycle of movement. Its speed is greatest at the time the tail fin is in line with the axis of the vertebral column and decreases on either side of the axis.

The maximum speed at which fish can swim varies from $6\frac{1}{2}$ miles per hour for the trout to as much as 27 miles per hour for the barracuda, which is the fastest fish known. Mammals, such as whales and dolphins, have been recorded swimming 20 and 22 miles per hour, respectively. Studies pertaining to large-sized mammals have indicated that their observed speed performance cannot be explained satisfactorily by ordinary hydrodynamic principles. *These animals are not capable of generating the power that has been calculated to be necessary to achieve the measured speeds.*

It has been suggested that whales and dolphins move much more efficiently than engineering calculations would suggest because they are able to reduce the turbu-

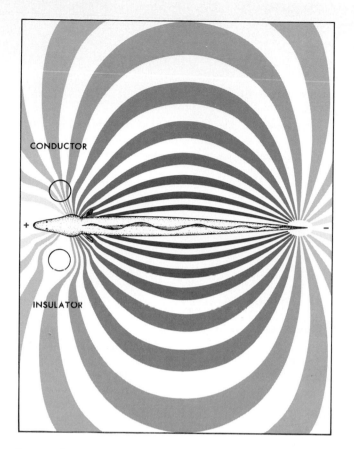

Figure 10-4 The *Glymnarchus niloticus'* detection system.

lence of water flow over their bodies. One possibility of accomplishing this is that the continual bending of the body, combined with the nonrigid nature of the skin and underlying muscles, reduces the resistance due to turbulence. Further study of this condition might produce results that could be adapted to the design of undersea craft.

An interesting variation of purposeful movement is found in the action of the wings of some insects, which beat so rapidly that they become invisible to the human eye. Insect wings are thin sheets of cuticle strengthened by hollow ribs. The wings do not contain muscles but are moved by the action of two sets of muscles attached to the walls of the thorax. By changing the dimensions of the thorax, the wings are moved up and down at the hinge joint where they are attached. In this way, the wings are able to move much more rapidly than if they were activated in the normal manner by muscle action.

Hydra, an animal belonging to the family Phylum Coelenterata, which live in the sea, has batteries of poisonous threads that may be shot out to capture prey. These thread capsules are adhesive and help to hold the prey. They are spiral shaped to entangle the prey better, and barbed to facilitate piercing the body of the prey. In all cases, they are found in the outer layer of cells of the body with a fine extension or trigger projecting from the surface. When the trigger is stimulated, the pressure within the capsule is increased suddenly and the coiled thread is expelled. The threads immobilize the prey, which can then be swallowed by the hydra. The exact propulsive mechanism by which the threads are expelled is not fully understood, but the propelling force is believed to come from a sudden increase in pressure.

TENTACLE

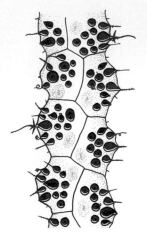

Figure 10-5 The hydra's tentacles are composed of batteries of thread capsules. In this drawing two of the large stinging capsules are discharged. The circles containing a central dot represent the top view of the capsules.

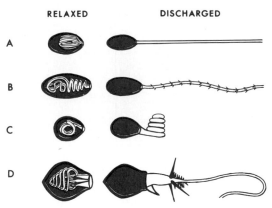

RELAXED DISCHARGED

A

B

C

D

Figure 10-6 Four types of thread capsules of the hydra. **A** and **B** are adhesive capsules, used to fasten the tentacles to solid objects when the hydra is looping. They also assist in capturing and holding the prey. **C** is a volvent type and aids in holding the prey by winding about it. **D** is a stinging capsule. It pierces the body of the prey and injects a poison.

Jellyfish belong to the same group of animals, and anyone who has brushed against or stepped on a jellyfish while swimming in the ocean is aware of the stinging potential of the protective capsules they possess. The Portugese man-of-war has a central disc-shaped body as large as 12 feet in diameter with numbers of tentacles over 100 feet in length. These tentacles contain stinging capsules, and swimmers becoming entangled in these tentacles have been rendered unconscious by the poison and have died.

Predatory fungi or molds have mechanisms to catch animals. Some molds have sticky filaments in which the prey becomes entangled; others form networks of loops that secrete a very sticky fluid in which its prey is held fast. Other predatory fungi form small rings of filaments that are just the diameter of its prey. When the prey sticks its head into one of these rings, it triggers an almost instantaneous reaction with the ring inflating and catching the prey. The mechanism of the reaction is not well understood, but it has been suggested that it results from the intake of water caused by a change in *osmotic* pressure, or from changes in the structure of the *colloids* that make up the *protoplasm* of the cells.

Perhaps the best-known plants with a specialized means of obtaining food are the carnivorous plants, which include the cobra plant, pitcher plant, Venus-flytrap, and the sundew.

Illustration 10-11

The Portugese man-of-war of the genus Physalia *is supported on the water by a large gas-filled float, which can be exhausted or filled at will. It is most often accompanied by a fish that lives among the tentacles of the hydroid and shares the food caught. As long as the fish is healthy and has no wounds, it is not harmed by the* Physalia. *However, once wounded it too falls prey to its host. The nature and origin of its normal immunity to the stinging action is not understood.*

Illustration 10-12

Many people consider meat-eating plants to be among the rarest of oddities, but there are over 450 species of plants that depend in part upon the food which they trap. The pitcher plant is such a carnivorous plant. Insects are lured over the edge or lip of the plant to reach nectar located just out of reach. Invariably, they fall into the abyss of the trap and into a watery soup designed to stupify, kill, and digest the insect. The inside walls of the trap have a very smooth surface, making it virtually impossible for the drenched insect to gain a footing.

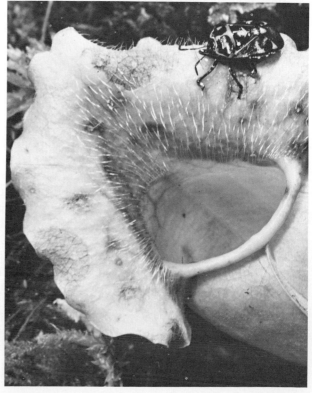

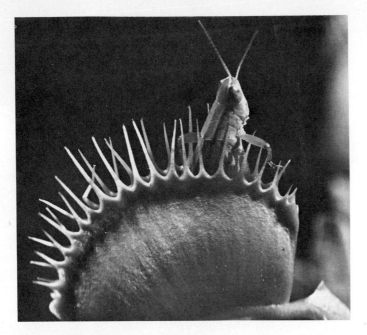

Illustration 10-13
The Venus' flytrap, found only in the southeastern part of the United States, was first discovered by Governor Arthur Dobbs of North Carolina, which he announced in a letter dated January 24, 1760. Each leaf of the plant is a small bear-trap-like mechanism consisting of two dished lobes that are effectively hinged along one side and armed with sharp spike teeth along the other edges. Within the inner surfaces are three tiny trigger hairs that, when touched ever so gently, cause the two lobes to snap shut and seal the doom of the insect prey within. After about 10 days the plant has assimilated the body tissues of the insect, and the jaws open ready for another catch. Although much effort has been expended in trying to solve the mechanism of the trap closure, there is as yet no definite answer. There is strong evidence, however, that the action is caused by electrical disturbances across the leaf of the trap.

The sundew plant has small red leaves less than half an inch across covered with tentacles that secrete a sticky substance for catching insects. Drops of this substance shine like dewdrops in the sunlight and are attractive to insects. When an insect touches one of the sticky tentacles, a reaction is triggered that results in the neighboring tentacles bending over the prey. The tentacles push the insect down onto the leaves of the plant where it is digested by the action of enzymes. The mechanism for the movement of the tentacles has attracted much interest because there are no muscles in plants to cause such movement. It has been determined that the movement is caused by growth phenomena. After the insect is digested, the tentacle unbends and is able to attract and catch another insect.

The Venus' flytrap is a small plant with leaves that are modified into two hinged lobes with sharp spines of teeth along the outer edge. Insects are attracted to the plant by secretions from glands located just inside the lips of the lobes. Each lobe has three small trigger hairs on its inner surface. Normally, the lobes are open until an insect enters and touches one of the trigger hairs. Then the lobes snap shut and the spines interlock to prevent the escape of the insect. The mechanism of snapping the lobes together is not well understood. Other glands on the inner surface of the lobes secrete the digestive enzymes that digest the insect. After the insect is digested, in perhaps 10 days, the lobes open again and are ready to capture another insect.

Special senses

The organs of the body concerned with the special senses of seeing, hearing, feeling, and communicating are among the most highly specialized structures in living things. Some of these specialized structures are of great interest to the engineer.

Most animal forms are sensitive to light. In some forms the sensitivity is a function of the entire body; in others there are specialized structures or regions of the body that perceive light stimuli. Some of these are no more than pigmented spots; others are detailed complicated structures that are highly specialized for their specific function. The mammalian eye is an example of a highly specialized structure.

Most eyes of higher animals are based on the general plan exemplified by the mammalian eye. However, some have interesting modifications that are adapted for the special purposes needed for the organism. One example is the eyes of chameleons, which are stalked and can be rotated in opposite directions. Such an adaptation aids the animal in capturing its prey. When the prey is in range, the chameleon shoots out its tongue (which is longer than its body) and picks up the prey on the enlarged sticky end of the tongue.

Illustration 10-14
Zot!

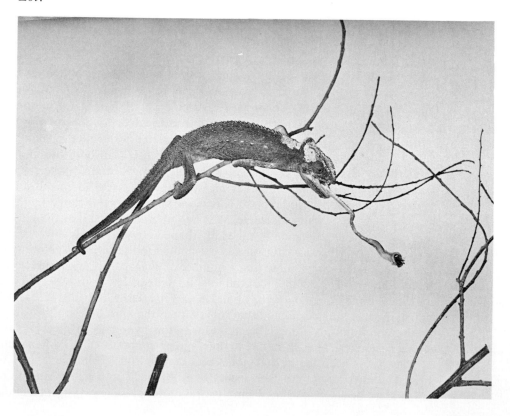

Time sense

One of man's preoccupations in today's busy, complicated world is the accurate determination of time. Star time, time clocks, tuning-fork timers, and electronic timers are all monuments to man's attempts to calibrate time. A time sense is part of all living things. Birds, bees, squirrels, insects, and man have many of their activities tightly regulated by "biological clocks." Such diverse activities as waking, feeding, storing and consuming oxygen, clotting blood, and singing have been associated with these biological clocks. The clocks seem to be independent of light and can be reset experimentally. Anyone who has traveled long distances east or west by jet aircraft is familiar with the fact that it takes a few days to adjust to the new time zone.

The exact mechanism of these biological clocks and where it is centered in the organisms has not yet been determined.

Shapes, aesthetics, and function

Many designs in nature might be justified on the basis of their beauty alone. However, investigation invariably reveals that also there is a highly functional reason for their structure. Some of these designs already have been adapted by man for his purposes. Others are still being studied but are not completely understood. As man learns more about their functions in nature, he may be able to adapt them to his purposes.

Diatoms are beautiful microscopic one-celled plants that exist in incredibly large numbers in both fresh and salt water. Each diatom secretes a shell of silicon around itself. Although the basic plan of these shells is the same, many of the 10,000 different species have very characteristic shapes, sizes, and structure. When alive, diatoms serve as a major food source for aquatic organisms. When they die, the shells fall to the bottom of the ocean, where they accumulate by hundreds of billions of tons. Many such deposits of diatomaceous earth from the floors of ancient seas are mined and used as insecticides, polishes, filters, insulation and as filler in paints, plastics, rubber, and roofing materials. Thus, diatoms are useful in both life and death.

Certainly comparable to diatoms in beauty, variety, and complexity are snowflakes. Although it is believed that no two snowflakes are identical, all are constructed on a basic hexagonal plan. This shape apparently is dictated by the molecular forces within the water molecule. Man has often wondered why these forces seem always to result in producing hexagonally shaped snowflakes. However, the production of this geometric configuration is not limited to the action of inanimate forces. Rather, it is interesting to note that a number of living organisms produce structures that are hexagonal in shape; for example, for years it has been observed that when honeybees build honeycomb, they build hexagonal cells with trihedral bases. Mathematicians have long postulated that this shape is the most efficient for the purposes honeycomb fulfills. Not everyone agrees with this hypothesis. However, the honeybee seems unconcerned about the matter, and he continues to build hexagonal cells.

Also it has been found that honeybees demand rather precise spatial relationships. The discovery of the "bee space" by Root made possible man-made beehives and modern beekeeping. Stated simply, bees prefer passageways $\frac{3}{16}$ inch in diameter.

Illustration 10-15

Diatoms are not sea gems or jewelry designs, but rather the exquisite little "glass" houses of single-celled aquatic plants so small that 15 million constitute a thimbleful. They live everywhere there is water; in every stream, lake, pond, and ocean. As living organisms they provide about 90 per cent of all the food in the ocean. In death their skeletons are used in a multitude of engineering applications.

Illustration 10-16

Honeybees build hexagonal cells with trihedral bases. Engineers, particularly those in the aircraft industry, have made good use of this configuration in designing wing and fuselage components that require a high strength-to-weight ratio. The elephant is standing on a lightweight panel made of hexagonal aluminum foil cells bonded between two thin aluminum sheets.

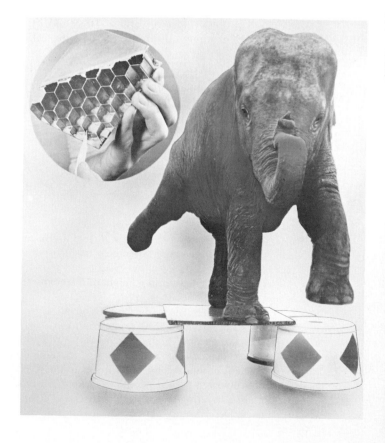

If smaller openings occur, they are plugged up by the bees with beeswax or plant products; larger openings are reduced in size by similar means to the preferred "bee space." For this reason all commercial beehives are designed to satisfy the bees' specifications.

The test of time

The structures and processes mentioned in this chapter are the present stage of design development that has been taking place over millions of years in living organisms. Countless variations of these designs have been tried and tested against the selective demands of the environment, but only those designs which conferred some advantage to the organism under existing environmental conditions were able to survive and be passed on to later generations. Each new generation of organisms contains variations of the basic design, which are then tested against the demands of the environment. Each new generation represents a "new model year" for living things.

This constant shifting, winnowing, and honing has resulted in organisms remarkably well adapted to the various enviromental conditions that they characteristically face. It should not be assumed, however, that the organisms that exist now are the end products of the perpetually changing process and that they will remain as they now are. In addition to the design variations already mentioned, in each generation of organisms the environmental conditions are also continually changing, as witness the current problems of pollution of air, water, soil, and in fact the entire biosphere.

As the environmental conditions change, different design variations will be selected. A design that was successful under one set of environmental conditions may not be capable of surviving under the changed conditions, whereas a previously unimportant design variation may make the difference between dying and living. In this way the design is constantly updated to existing conditions. If there is anything certain about living organisms, it is the fact that they are constantly undergoing change.

Problems

10-1. Pick a leaf from a broad-leaved tree and examine it carefully.
(a) What engineering principles are incorporated into this leaf to enable it to carry out its functions?
(b) Drop the leaf from a height of 6 feet and study its behavioral characteristics as it falls to the ground.
(c) What happens to the leaf if you put it in a pan of water?
(d) Compare the above results with those of a dry leaf of the same kind. Are the results the same? Explain.

10-2. Take a needle from a pine tree and examine it closely. These needles are said to be efficient in minimizing water losses to the plant. Enumerate the engineering principles involved for the needle to accomplish this.

10-3. Place a live frog on the desk before you. Observe his body and leg actions when he hops. Describe any physical or engineering principles involved in these movements.

10-4. Place a live frog in an aquarium or pan of water and carefully observe his actions and movements in swimming. Describe any physical or engineering principles involved in these movements.

10-5. Using your observations of a swimming or hopping frog, calculate the amount of work done per unit of time by each activity. Are the results the same for both activities? Explain.

10-6. Observe the activity of a dragonfly nymph in an aquarium. Notice particularly the various methods that he utilizes in locomotion. Try to determine his rate of movement for each method of locomotion.

10-7. Utilizing your observations of problem 10-6, calculate the forces involved and the amount of work done in each of the observed activities. Compare the results.

10-8. Obtain a 12-inch length of bamboo and a similar length of two or more other kinds of wood of the same diameter as the bamboo. Compare them in regard to weight, flotation, water displacement, and breaking strength. How are these characteristics related to the structure of each of the pieces of wood?

10-9. Given a live cockroach, a piece of string, some glue, and a strobe light, how would you go about determining the frequency of wing beat of the cockroach when it is flying?

10-10. Using sound as a measurement, how could you determine the frequency of the wing beat of a housefly?

10-11. Examine the wrist bones and joints of a human skeleton. Considering the functions that the wrist performs, how could its design be improved? Explain.

10-12. Examine a leg bone or wing bone of a chicken. Cut or break it open and examine its internal structure. What are the reasons for its strength-to-weight ratio?

10-13. Watch a dog or horse trot. What is the sequence of movement of the legs? How long does it take for one cycle of leg movements?

10-14. The statement has been made that a dog or horse trotting across a bridge has a greater chance of causing the bridge to collapse than does a much heavier object rolling across the bridge. Is there any basis for this claim? Explain.

10-15. What advantages or disadvantages does a four-footed animal have as compared to man in regard to
(a) Jumping
(b) Running
(c) Swimming
How important are the mechanical characteristics of the skeleton?

10-16. Obtain a package of seeds (squash, watermelon, sunflower) from the grocery store and examine them carefully. What functions are performed by seeds? How is their structure adapted to carry out these functions?

10-17. Examine the outside of an orange. What is the purpose of the apparent pores or depressions? Carefully peel the orange. Do the pores go all the way through the peel? What functions does the orange peel perform? How are these functions related to its structure?

10-18. Carefully separate the individual sections of an orange from one another. Notice their shape, structure, and the manner in which they are held together? Can you think of any design advantages to such a shape?

10-19. Examine the general shape of a grape, an orange, a tomato, and an apple. As far as plants are concerned, what functions are carried out by each? Of what value are their general shapes in carrying out these functions? Are there any special structural modifications that also help with these functions?

10-20. In "Indian wrestling" the two contestants place their elbows on the table and attempt to force their opponent's wrist down onto the table. Draw a force diagram of such a contest and indicate whether mechanically the advantage would be to the person with a long or a short elbow-to-wrist distance.

10-21. Some animals are built around a spherical body plan (like a sponge); others are built on an elongated body plan (like a fish). In general, animals with a spherical body plan are slow moving and sedentary and those with an elongated body plan are faster moving. What features of the two shapes might contribute to their movement capability?

10-22. Examine the structure of a feather. What functions do feathers perform for birds? How is the feather structure related to these functions?

10-23. Study the shape of any large tree. Does there seem to be any particular arrangement to the location of the branches? Is this arrangement true also for other kinds of trees? Describe your general observations. What is the advantage to the plant of such an arrangement—or lack of one?

10-24. One of the important functions of green leaves is to carry on food manufacture through the process of photosynthesis. Light is required for this process. Study the general shape and structure of a large tree. Can you detect any evidence of a positioning of branches and leaves that might be related to the function of food manufacture? *Explain.*

10-25. Examine a chicken egg. From a chicken's standpoint what is the function of an egg? Break the egg open. How is the external and internal structure of the egg related to its function?

10-26. Carefully wax half of a smooth surface. Sprinkle water on both halves and observe the shape of the water droplets. What shape do most of the droplets seem to have on the waxed portion of the surface? Are the droplets the same shape on the unwaxed portion? Explain the differences, if any, in the shapes.

10-27. Examine a human hair. What are some of the functions of hair to animals? How are these functions related to the general structure of a hair?

10-28. What is the function of hair in one's armpits? How else might this same function be accomplished?

10-29. Can people with big ears hear better than those with small ears? What is the major function of the external ear? How does its design accomplish this function?

10-30. Name some mechanical devices for receiving sound in which the principles used by the external ears of animals have been adapted and employed.

10-31. Observe live fish in an aquarium. Describe the procedure whereby they swim through the water.

10-32. How would you proceed to determine the forces that are involved when a fish propels itself through the water?

10-33. Examine an unpainted piece of wood. What causes *grain* in wood? Of what value is grain to a tree?

10-34. Pull up a plant, being careful to get as much of the root system as possible. Carefully examine the roots, stem, and leaves and attempt to relate the design structure of each to its function for the plant.

10-35. Examine carefully the structure of a flower. Consider its main function to the plant to be that of reproduction. To accomplish this, it is most likely that insects must be attracted to the flower. Relate the structure of the various parts of the flower to these functions.

11

The design process

To many people engineering design means the making of engineering drawings, putting on paper ideas that have been developed by others, and perhaps supervising the construction of a working model. While engineers should possess the capability to do these things, the process of engineering design includes much more: the *formulation* of problems, the *development* of ideas, their *evaluation* through the use of models and analysis, the *testing* of the models, and the *description* of the design and its function in proposals and reports.

An engineering problem may appear in any size or complexity. It may be so small that an engineer can complete it in one day or so large that it will take a team of engineers many years to complete. It may call for the design of a tiny gear in a big machine, perhaps the whole machine, or an entire plant or process which would include the machine as one of its components. When the design project gets so big that its individual components can no longer be stored in one person's head, then special techniques are required to catalog all the details and to ensure that the components of the system work harmoniously as a coherent unit. The techniques which have been developed to ensure such coordination are called *systems design*.

Regardless of the complexity of a problem that might arise, the *method* for solving it follows a pattern similar to that represented in Illustration 11-1. Each part of this "cyclic" process will be described in more detail, but first, two general characteristics of the process should be recognized.

1. Although the process conventionally moves in a circular direction, there is continuous "feedback" within the cycle.
2. The method of solution is a repetitious process that may be continuously refined through any desired number of cycles.

The concept of *feedback* is not new. For example, feedback is used by an individ-

Illustration 11-1
The design process begins with a need and gathers definition in a continuous evolutionary spiral.

ual to evaluate the results of actions that have been taken. The eye sees something bright that appears desirable and the brain sends a command to the hand and fingers to grasp it. However, if the bright object is also hot to the touch, the nerves in the fingers feed back information to the brain with the message that contact with this object will be injurious, and pain is registered to emphasize this fact. The brain reacts to this new information and sends another command to the fingers to release contact with the object. Upon completion of the feedback loop, the fingers release the object.

Another example is a thermostat. As part of a heating or cooling system, it is a feedback device. Changing temperature conditions produce a response from the thermostat to alter the heating or cooling rate (Figure 11-1).

The rate at which one proceeds through the problem-solving cycle is a function of many factors, and these factors change with each problem. Considerable time or very little time may be spent at any point within the cycle, depending upon the situation.

Thus, the problem-solving process is a dynamic and constantly changing process that provides allowances for the individuality and capability of the user.

The *design process* is used in each of the phases of design that are described in Chapter 12. Each phase starts with the identification of the problem and ends with a report. Some parts of the loop are more important in one phase than another. For example, the search for ideas is most important during the feasibility study and the preliminary design phases, as compared to analysis and experimentation, which tend to predominate in the preliminary and detail design phases. The *solution* of one phase often leads directly to the problem formulation for the next phase.

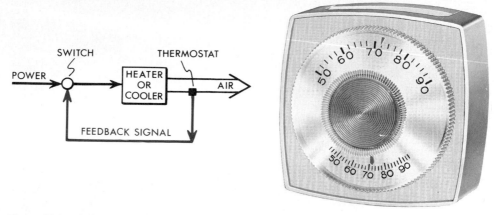

Figure 11-1 The function of a thermostat can be represented diagramatically.

Identification of the problem

One of the biggest surprises that awaits the newly graduated engineer is the discovery that there is a significant difference between the classroom problems solved in school and the real-life problems that he or she is now asked to solve. This is true because problems encountered in real life are poorly defined (Illustration 11-2). The individuals who propose such problems (whether they be commercial clients or the engineer's employer) rarely know or specify exactly what is wanted, and engineers must decide for themselves what information they need to secure in order to solve the problem. In the classroom you will be confronted with well-defined problems, and you usually will be given most of the facts necessary to solve them in the problem statements. In a real situation you will find that you have available insufficient data in some areas and an overabundance of data in others. In short, you must first find out what the problem *really* is. In this sense you will be no different from the physician who must diagnose an illness or the attorney who must research a case before appearing in court. In fact, problem formulation is one of the most interesting and difficult tasks that the engineer faces. It is a necessary task, for one can arrive at a good and satisfactory solution only if the problem is fully understood. Many poor designs are the result of inadequate problem statements.

The ideal clients who hire a designer to solve a problem will know what they want the designer to accomplish; that is, they know their problem. They will set up a list of limitations or restrictions that must be observed by the designer. They will understand that an *absolute* design rarely exists—a *yes* or *no* type of situation—and that the designer usually has a number of choices available. Clients can specify the most appropriate optimization criteria on which the final selection (among these choices) should be based. These criteria might be cost, or reliability, or beauty, or any of a number of other desirable results.

As an engineer you must determine many other basic components of the problem statement for yourself. You must understand not only the task that the design is required to perform but what its range of performance characteristics is, how long it is expected to last in the job, and what demands will be placed on it one year, two years, or five years in the future. You must know the kind of an environment in which the design is to operate. Does it operate continuously or intermittently? Is it

Illustration 11-2
In the design process one of the most difficult tasks is to define accurately the problem.

subject to high temperatures, or moisture, or corrosive chemicals? Does it create noise or fumes? Does it vibrate? In short, what type of design is best suited for the job?

For example, let us assume that as an engineer you have been asked by a physician to design a flow meter for blood. What do you need to know before you can begin your design? Of course you should know the quantity of blood flow that will be involved. Does the physician want to measure the flow in a vein or in an artery? Is it important to measure the flow in the very small blood vessels near the skin or rather in the major blood vessels leading to and from the heart? Do you want to measure the average flow of blood or the way in which the blood flow varies with every pulse beat? How easy will it be to have access to the blood vessels to be tested? Will it be better to measure the blood flow without entering the vessel itself, or should a device be inserted directly into the vessel? One major problem in inserting any kind of material into the blood stream is a strong tendency to produce blood clots. In case an instrument can be inserted into the vessel, how small must it be so that it does not disturb the flow which it is to measure? How long a section of blood vessel is available, and how does the diameter of the blood vessel vary along its length and during the measurement? These and many more components of the problem statement must be determined by you before an effective solution can be designed (Figure 11-2).

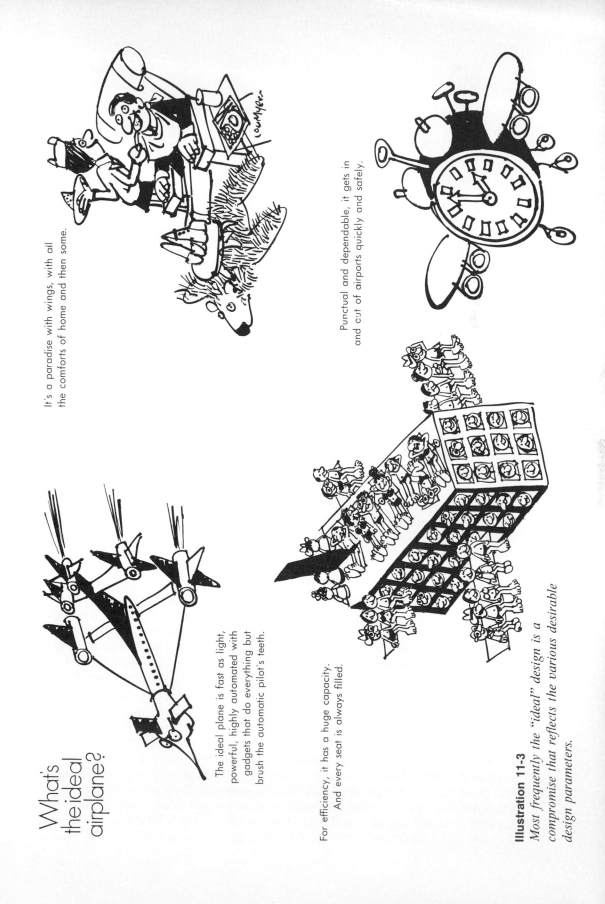

It's a paradise with wings, with all the comforts of home and then some.

Punctual and dependable, it gets in and out of airports quickly and safely.

What's the ideal airplane?

The ideal plane is fast as light, powerful, highly automated with gadgets that do everything but brush the automatic pilot's teeth.

For efficiency, it has a huge capacity. And every seat is always filled.

Illustration 11-3
Most frequently the "ideal" design is a compromise that reflects the various desirable design parameters.

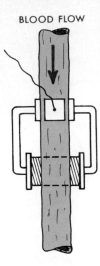

BLOOD FLOW

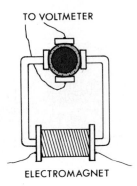

TO VOLTMETER

ELECTROMAGNET

Figure 11-2 A conceptual model of an electromagnetic blood flow meter.

Another example of the importance and difficulty of problem definition is the urban transportation problem. Designers have proposed bigger and faster subways, monorails, and other technical devices because the problem was assumed to be simply one of transporting people faster from the suburbs into the city. In many cases, it was not questioned whether the problem that they were solving was *really* the problem that needed a solution.

Surely, the suburbanite needs a rapid transportation system to get into the city, but the rapid transport train is not enough. He must also have "short haul" devices to take him from the train to his home or to his work with a minimum of walking and delay. Consequently, the typical rapid transit system must be coordinated with a citywide network of slower and shorter-distance transportation which permit the traveler to exit near his job, wherever it may be. For the suburbanite, speed is not nearly as important as frequent, convenient service on which he can rely and for which he need not wait.

Urbanites, particularly the poor, who generally live far from the places where they might find work, are also in need of better transportation. For these people, high speed again is not nearly as important as low cost and transportation routes and vehicles that provide access to the job market. Instead of placing emphasis on bigger and faster trains, designers should consider the *wants* and the *needs* of the people they are trying to serve and determine what these wants and needs really are.

How does the engineer find out? How can the problem be defined so that its solution will answer the *real* need? Of course the first step is to find out what is

Illustration 11-4

The design of rapid transportation systems of the future must be carefully coordinated with the design of the city.

already known. The literature must be studied. The engineer must become thoroughly familiar with the problem, with the environments in which it operates, with similar machines or devices built elsewhere, and with peculiarities of the situation and the operators. *The engineer must ask questions.*

It may be, after evaluating the available information, that the engineer will be convinced that the problem statement is unsatisfactory—just as today's statement of the transportation problem appears to be unsatisfactory. In that case he may suggest or perform additional studies—studies that involve the formulation of simulation models of the situation and the environment in which the machine is to be built. They may include experiments with these models to show how this environment would react to various solutions of the problem.

The design engineer must work with many types of people. Some will be knowledgeable in engineering—others will not. His design considerations will involve many areas other than engineering, particularly during problem formulation. He must learn to work with physicists and physicians, with artists, architects, and city planners, with economists and sociologists—in short, with all those who may contribute useful information to a problem. He will find that these men have a technical vocabulary different from his. They look at the world through different eyes and approach the solution of problems in a different way. It is important for the engineer to have the experience of working with such people before he accepts a position in industry, and what better opportunity is there than to make their acquaintance during his college

Illustration 11-5
The engineer's work is frequently worldwide in its scope. Consequently, he or she must be adaptive to new and unusual physical environments and be able to work with many types of people.

years. With the manifold problems that tomorrow's engineer will face—problems that involve human values as well as purely technical values—collaboration between the engineer and other professional people becomes increasingly important.

Collection of information

The amount of technical information available to today's scientists and engineers is prodigious and increasing daily. Two hundred years ago, during the time of Jefferson and Franklin, it was possible for an individual to have a fair grounding in all the social and physical sciences then known, including geography, history, medicine, physics, and chemistry, and to be an authority in several of these. Since the Industrial Revolution, or about the middle of the last century, the amount of knowledge in all the sciences has grown at such a rapid rate that no one can keep fully abreast of one major field, let alone more than one. It has been estimated that if a person, trained in speed reading, devoted 20 hours a day, seven days a week to nothing but study of the literature in a relatively specialized field, such as mechanical engineering, he would barely keep up with the current literature. He would not have time to go backward in time to study what has been published before or to consider developments in other fields of engineering. How then can one be able to find information that is available, or know what has been done concerning the solution of a particular problem? The answer is twofold: know *where* the information resources are located, and know *how* to retrieve information from a vast resource.

A typical technical library may contain from 10,000 to 200,000 books. It may subscribe to as many as 500 technical and scientific magazines, as well as a large store of technical reports published at irregular intervals by government agencies, universities, research institutes, and industrial organizations. The problem then is principally one of finding the proper books or articles.

Libraries have become quite efficient at cataloging books and major reports in their general catalog file. Usually these catalogs are arranged into three groups, one by author, one by title, and one by subject matter. Although the library catalog is an excellent source of book references, it does not contain any of the thousands of articles in magazines, technical journals, and special reports.

One's direction to these journal articles and special reports is through the reference section in the library. Here the abstract journals and books devoted to collecting and ordering all publications in a particular field are housed. For engineers, two of the most useful of these are the *Engineering Index* and the *Applied Science and Technology Index*. They appear annually and contain short abstracts of most important articles appearing in the engineering field. Articles are organized according to subject headings, so that all articles on a similar subject appear together. By looking under the appropriate heading, the searcher can discover the references of greatest interest to him or related headings where other references might be found. After satisfying himself that he has the correct references, the researcher then goes to the appropriate periodicals to find the full articles. There are many other indexes besides the two mentioned above, some more, some less specialized.

As an example, assume that we are concerned with the design of a pipeline to transport solid refuse (garbage) from the center of a large city to a disposal site where it may be processed, incinerated, or buried. Let us follow and observe an engineer making the required library search.

Literature search on "pipelining of refuse"

(The following "capsule narrative" indicates what actually happened during a quick

Illustration 11-6
An overabundance of data does not necessarily guarantee that the engineer's task will be simplified.

Illustration 11-7

Computerized retrieval systems utilizing millions of pages of data are a significant improvement for the engineer over more traditional library systems.

noncomprehensive search conducted in an afternoon at a typical university library, and illustrates the kind of search that an engineer might make for a brief study.)

Going first to the "subject" section of the catalog file, I could think of only three headings to look under: Refuse, Garbage, and Pipelines. There were 24 entries under "Refuse and Refuse Disposal." Some dealt with conveyors and trucking but only one with pipelining (not surprisingly since this is not a common way to convey garbage). I copied some of the titles and reference numbers because they might help to give me some idea of the composition and consistency of garbage and of the shredders and other devices used to make refuse more uniform in size and more capable of being transported in a pipeline.

The card under "Garbage" referred me right back to "Refuse and Refuse Disposal," a dead end.

Under "Pipe" there were some 75 entries under 23 different subheadings from "Pipe-Asbestos, Cement" to "Pipe-Welding." I copied the titles and numbers shown in Illustration 11-8.

I wasn't satisfied that I had exhausted the subject file but could not think of any other pertinent headings. So I went next to the reference library and sat down in front of the shelf with the *Engineering Index*. The latest complete year was 1984. Looking under the headings like "Pipeline" and "Refuse Disposal", I paid particular attention to "See also" lists (Illustration 11-9) and these eventually led me to a veritable gold mine of references under "Materials Handling." A few of these are shown in Illustration 11-10. Notice that two of the most interesting articles are in German. If I find them, I will have to have them translated.

Fool me once, shame on you; fool me twice, shame on me.
—Chinese proverb

Illustration 11-8

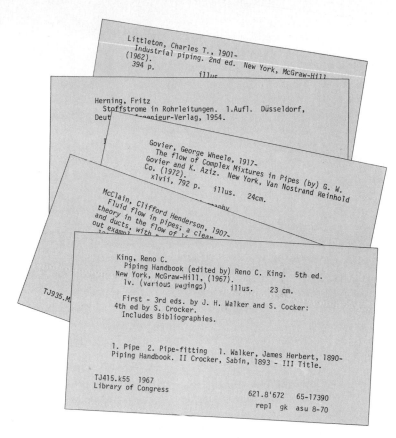

Now that I know some of the best headings, I look through other years' editions of the *Index* and also study other indexes. I can also go back to the subject catalog and look under the headings that have been productive in the *Index,* headings like "Materials Handling," of which I did not think the first time. In this way, in a short afternoon, one can assemble a reasonably good reference list on any subject one needs to study.

Next I have to obtain copies of those articles that I want to study in their entirety. If I found them as books in the subject catalog, I can ask for them or look for them myself in the stacks. For articles which have appeared in magazines, like the references from the *Engineering Index,* I first have to find out whether these references are available in the library and, if so, I'll find the appropriate order number and borrow them from the library. If they are particularly interesting, I may have the library make me a photostatic copy so that I can have a permanent record of the article.

If it is important for the searcher to find the very latest work done in his subject, the library will not be of much help. There is a time delay between the performance of a piece of research, its publication, and its appearance in any of the abstract

---

Knowledge is of two kinds: we know a subject ourselves, or we know where we can find information upon it.
—Samuel Johnson

Illustration 11-9
Typical headings from Engineering Index.

REFUSE DISPOSAL

See Also Coal Mines and Mining–Waste Disposal; Coal Preparation–Refuse Disposal; Industrial Wastes.

TRANSPORTATION

See also Air Transportation; City Planning; Coal Transportation; Highway Engineering–Research; Highway Systems–Planning; Industrial Plants–Location; Mines and Mining–Surface Transportation; Motor Bus Transportation; Motor Truck Transportation; Noise–Control; Operations Research; Power Plant Equipment–Transportation; Railroads; Rapid Transit; Street Traffic Control; Traffic Surveys; Vehicles–Ground Effect; Waterway Transportation.

COAL HANDLING

See also Bins; Coal Transportation.

Hydraulic. See also Materials Handling–Pipe Lines; Ore Handling–Hydraulic; Pipe, Plastic.

MATERIALS HANDLING

See also Air Transportation; Aircraft Carriers; Barges; Bins; Bottle Caps; Cableways; Cargo Handling; Cement Handling; Chemicals–Transportation; Coal Handling; Construction Equipment; Containers; Conveyors; Cranes; Elevators; Flour Mills; Food Products–Packaging; Foundry Practice–Cleaning; Freight Handling; Furniture Manufacture–Finishing; Grinding Mills; Hoists; Industrial Engineering; Industrial Trucks; Logging; Machine Tools; Mail Handling; Mines and Mining–Surface Transportation; Nuclear Fuels–Safe Handling; Ore Handling; Packaging; Packaging Machines; Packaging Materials; Production Planning and Control; Rivers–Revetments; Salt–Storage; Scales and Weighing; Silos; Systems Engineering; Transportation; Warehouses.

Illustration 11-10
Typical abstracts from Engineering Index.

042824 UMSCHLAG VON MASSENGUETERN SYSTEMATIC DER ANLAGEN-KOMPONENTEN. (Handling Bulk Goods, Systematics of the Installation Components.) Handling bulk goods is defined in all the various forms starting from the purpose of handling and storage. It is always resorted to where goods have to be transferred from one handling equipment to the other. The author describes the various methods of transshipment of bulk goods which can be classified according to the method of accumulation or to the time the goods remain in the rail and water termina. He also deals with the assessment of transshipment and distribution equipment and bulk goods transshipment installations in harbors. 10 refs. In German with English abstract.

Lubrich, W. S. W. Foerdern Heben v 23 n 14 Oct 1973 p 763-766.

042828 MATERIAL HANDLING AND THE INDUSTRIAL ENGINEER. The subjects discussed at this conference held in Philadelphia, Pa. from Jan 16-18 1974 can be broadly categorized as follows: analysis of materials handling problems which is a vital first step to efficient operation, selection criteria and determination of costs, computers in materials handling, automatic warehousing and retrieval systems and efficient plant layout and planned material flow. The 6 conference papers are indexed separately.

Semin, AIIE Mater Handl and the Ind Eng. Semin. 3rd Tech Pap. Philadelphia, Pa. Jan 16-18 1974 Publ by AIIE, New York, 1974, 44p.

042939 PROPOSED TRANS-ALASKA INTEGRATED PIPELINE TRANSPORTATION SYSTEM. The article describes the proposed Integrated Pipeline Transportation (IPT) system across Alaska (including its structural design features) and how it could be constructed, gives the advantages and obstacles to the plan, a breakdown of the cost, and discusses how the project could be financed.

Nasser, George D. Prestressed Concr Inst. Chicago, Ill. J Prestressed Concr Inst v 18 n Sep-Oct 1973 p 18-31.

042940 UEBERBLICH UEBER BETRIEBSAN-LAGEN ZUM HYDRAULISCHEN FESTSTOFF-TRANSPORT. (Survey of Installations for Pipelining of Solids.) Paper contains a world-wide survey of the most important installations for pipelining of solid materials with particular emphasis on installations located in Eastern Europe and Asia. Characteristic data related; throughput (slurry and solids); velocity of the flow, diameter of the pipeline; specific power requirements; screen analysis of the solids; distance involved; and type of solids are given in tabulated form. 21 refs. In German.

Boehme, Frank Ration Braunkohle, Grossaeschen, E Ger. Neue Bergbautech v 3n 8 Aug 1973 p 567-570.

061680 SIZE CHARACTERISTICS OF MUNICIPAL SOLID WASTE. A reclamation system currently under development at MIT seeks to separate automatically many items of refuse in the form in which they come from the packer truck. This system eliminates the unnecessary cost of pulverizing or pulping homogeneous refuse items such as bottles, cans and newspapers, and at the same time preserves them in the form in which their recycle value is a maximum. 5 refs.

journals. This delay is usually three years or more. The only source for the very latest materials is the expert himself. If one has made an exhaustive literature survey, he has usually found one or more researchers who are specialists and have published extensively in the field under investigation. These people are also the ones who probably can provide the latest technical information in the field. Often these men are happy to share their knowledge with the searcher in the field. However, it is customary to offer them a consultant's remuneration if a substantial amount of their time is required for this service.

Generation of ideas

As man counts time, the first act of recorded history was one of creation. When God created man, he endowed him with some of this ability to bring new things into being. Today the ability to think creatively is one of the most important assets that all men possess. The accelerated pace of today's technology emphasizes the need for conscious and directed imagination and creative behavior in the engineer's daily routine. However, this idea is not new.

For centuries primitive man fulfilled his natural needs by using the bounty nature placed about him. Since his choice was limited by terrain, climate, and accessibility, he was forced to choose from his environment those things which he could readily adapt to his needs. His only guide—trial and error—was a stern teacher. He ate whatever stimulated his sense of smell and taste, and he clothed himself and his family in whatever crude materials he could fashion to achieve warmth, comfort, and modesty. His mistakes often bore serious consequences, and he eventually learned that his survival depended upon his ability to think and to act in accordance with a plan. He learned the importance of imaginative reasoning in the improvement of his lot.

In recent years archaeologists have discovered evidence of early civilizations that made hunting weapons and agricultural tools, mastered the use of fire, and impro-

Illustration 11-11
"In the beginning God created the heavens and the earth."
(Genesis 1:1.)

vised fishing equipment from materials at hand—all at an advanced level of complexity. These remains are silent reminders of man's ingenuity. Only his cunning and imagination protected him from his natural enemies. The situation is much the same even today, centuries later. Many believe that in this respect man may not have improved his lot substantially over the centuries. The well-being of our civilization still depends upon how successfully we can mobilize our creative manpower. As a profession, engineering must rise to meet this challenge.

In scientific work the term *creativity* is often used interchangeably with *innovation*. However, the two are not synonymous although they do have some similarities. Both creativity and innovation refer to certain processes within an individual or system. Innovation is the discovery of a new, novel, or unusual idea or product by the application of logic, experience, or artistry. This would include the recombination of things or ideas already known. Creativity is the origination of a concept in response to a human need—a solution that is both satisfying and innovative. It is reserved for those individuals who originate, make, or cause something to come into existence for the first time or those who originate new principles. Innovation, on the other hand, may or may not respond to a human need, and it may or may not be valuable. In effect, creativity is innovation to meet a need.

It is incorrect to use the terms *synthesis, innovation,* and *creativity* interchangeably. They are not synonymous but all are used in the generation of ideas to solve engineering problems. *Synthesis* is the assembly of well-known components and parts to form a solution. *Innovation* is the discovery of a new, novel, or unusual idea or product by applying logic, experience, or artistry. *Creativity* originates an entirely new concept in response to a human need, a solution which is both satisfying and innovative. It presupposes an understanding of human experience and human values.

Problem solving does not necessarily require creative thought. Many kinds of problems can be solved by careful discriminating logic. An electronic computer can be programmed to perform synthesis—and perhaps even a certain degree of innovation—but it cannot create.

Creativity is a human endeavor. It presupposes an understanding of human experience and human values, and it is without doubt one of the highest forms of mental activity. In addition to requiring innovation, creative behavior requires a peculiar insight that is set into action by a vivid but purposeful imagination—seemingly the result of a divine inspiration that some often call a "spark of genius." Indeed, the moment of inspiration is somewhat analogous to an electrical capacitor that has "soaked up" an electrical charge and then discharges it in a single instant. To

---

Imagination is more important than knowledge.
—Albert Einstein

More today than yesterday and more tomorrow than today, the survival of people and their institutions depends upon innovation.
—Jack Morton, *Innovation,* 1969

The age is running mad after innovation. All the business of the world is to be done in a new way. Men are to be hanged in a new way.
—Samuel Johnson, 1777

Illustration 11-12

Engineers are motivated to work at a particular task partly because of the exhilaration, thrill, special satisfaction, pride, and pleasure they get from completing a creative task. However, the path to success is not always easy, and on occasion frustration is rampant.

sustain creative thought over a period of time requires a large reservoir of innovations from which to feed. Creative thought may be expressed in such diverse things as a suspension bridge, a musical composition, a poem, a painting, or a new type of machine or process. Problem solving, as such, does not necessarily require creative thought, because many kinds of problems can be solved by careful, discriminating logic.

The engineer who redesigns a radio or improves an automobile engine uses established techniques and components; he synthesizes. Innovators are those who build something new, and who combine different ideas and facts with a purpose. Creativity is one of the rarest and highest forms of human activity. We only call those individuals "creative" who originate, make, or cause to come into existence an entirely new concept or principle. (Patents are mostly the result of clever innovation, rather than creative effort.) If we had to rely on creativity for patents, we would not have over 4 million patents in the United States alone. All engineers must synthesize, some will innovate, but only a very few are able to be truly creative.

Years ago most American youths were accustomed to using innovative and imaginative design to solve their daily problems. Home life was largely one of rural experience. If tools or materials were not available, they quickly improvised some other scheme to accomplish the desired task. Most people literally "lived by their

Illustration 11-13
All persons of normal intelligence possess some ability to think creatively and to engage in imaginative and innovative effort.

wits." Often it was not convenient, or even possible, to "go to town" to buy a clamp or some other standard device. Innovation was, in many cases, "the only way out." A visit to a typical midwestern farm or western ranch today, or to Peace Corps workers overseas, will show that these innovative and creative processes are still at work. However, today most American youths are city or suburb dwellers who do not have many opportunities to solve real physical problems with novel ideas.

A person is not born with either a creative or noncreative mind, although some are fortunate enough to have exceptionally alert minds that literally feed on new experiences. Intellect is essential, but it is not a golden key to success in creative thinking. Intellectual capacity certainly sets the upper limits of one's innovative and creative ability; but nevertheless, motivation and environmental opportunities determine whether or not a person reaches this limit. Surprisingly, students with high IQs are not necessarily inclined to be creative. Recent studies have shown that over 70 per cent of the most creative students do not rank in the upper 20 per cent of their class on traditional IQ measures.

Everyone has some innovative or creative ability. For the average person, due to inactivity or conformity, this ability has probably been retarded since childhood. If we bind our hand or foot (as was practiced in some parts of the Orient) and do not use it, it soon becomes paralyzed and ineffective. But unlike the hand or foot, which cannot recover full usefulness after long inactivity, the dormant instinct to think creatively may be revived through exercise and stimulated into activity after years of

An inventor is simply a fellow who doesn't take his education too seriously.
—Charles F. Kettering

Everybody is ignorant, only in different subjects.
—Will Rogers

near suspended animation. Thus everyone can benefit from studying the creative and innovative processes and the psychological factors related to them.

Since there is always a great demand for creative and innovative ideas, many attempts have been made to develop procedures for stimulating them. Certain of these procedures will work satisfactorily in one situation, yet at other times different methods may be needed.

There are many methods of stimulating ideas that are used in industry today: (*a*) the use of checklists and attribute lists, (*b*) reviewing of properties and alternatives, (*c*) systematically searching design parameters, (*d*) brainstorming, and (*e*) synectics. These methods will be discussed briefly.

Illustration 11-14
The mind can die from inactivity.

Checklists and attribute lists

One of the simplest ways for an individual to originate a number of new ideas in a minimum amount of time is to make use of prepared lists of general questions to apply to the problem under consideration. A typical list of such questions might be the following:

1. In what ways can the idea be improved in quality, performance, and appearance?
2. To what other uses can the idea be put? Can it be modified, enlarged, or minified?
3. Can some other idea be substituted? Can it be combined with another idea?
4. What are the idea's advantages and disadvantages? Can the disadvantages be overcome? Can the advantages be improved?
5. What is the particular scientific basis for the idea? Are there other scientific bases that might work equally well?
6. What are the least desirable features of the idea? The most desirable?

Attribute listing is a technique of idea stimulation that has been most effective in improving tangible things—such as products. It is based upon the assumption that most ideas are merely extensions or combinations of previously recognized observations. Attribute listing involves the following:

1. Listing the key elements or parts of the product.
2. Listing the main features, qualities, or significant attributes of the product and of each of its key elements or parts.
3. Systematically modifying, changing, or eliminating each feature, quality, or attribute so that the *original purpose* is better satisfied, or perhaps a new need is fulfilled.

With both checklists and attribute lists one must be careful to recognize that these methods are merely "stimulators" and that they are not intended to replace original and intelligent thinking. They are certainly not intended to be used as crutches. Rather, like a wrench which extends the power or leverage of a man's fingers or arm, these ideation tools extend the power and effectiveness of the mind.

Reviewing the properties and alternatives

Another rather common procedure, somewhat similar to attribute listing, is to consider how all the various properties or qualities of a particular design might be changed, modified, or eliminated. This method lists the modifiable properties such as weight, size, color, odor, taste, shape, and texture. Functions that are desirable for the item's intended use may also be listed: automatic, strong, durable, or lightweight. After developing these lists, the engineer can consider and modify each property or function individually.

Imagine redesigning a lawn mower. The listed properties might include (1) metal,

(2) two cycle, gasoline-powered, (3) four wheels, (4) rotary blade, (5) medium weight, (6) manually propelled, (7) chain driven, and (8) green in color. In beginning the design of an improved lawn mower, the engineer might first consider other possibilities for each property. What other materials could be used? Can the engine be improved—what about using electrical power? Should the mower operate automatically? Should the type of blade motion be changed? Questions like these may suggest how the design *could* be improved. The properties of lawn mowers have been changed many times, and these changes have presumably made lawn mowers more efficient and easier to use.

Again, besides considering the product's various properties the engineer must question, observe, and associate its functions. Can these functions be modified, rearranged, or combined? Can the product serve other functions or be adapted to other uses? Can we change the shape (magnify or minify parts of the design)? With this type of questioning we can stimulate ideas that will bring design improvements to the product.

Systematic search of design parameters

Frequently it is advisable to investigate alternatives more thoroughly. A systematic search considers all possible combinations of given conditions or design parameters. This type of search is frequently called a "matrix analysis." Its success in stimulating ideas depends upon the engineer's ability to identify the significant parameters that affect the design. The necessary steps for implementing this type of idea search are the following:

1. **Describe the problem.** This description should be broad and general, so that it will not exclude possible solutions.
2. **Select the major independent-variable conditions** required in combination to describe the characteristics and functions of the problem under consideration.
3. **List the alternative methods** that satisfy each of the independent-variable conditions selected.
4. **Establish a matrix** with each of the independent-variable conditions as one axis of a rectangular array. Where more than three conditions are shown, the display can be presented in parallel columns.

Let us consider a specific example to see how this method can be applied.
 1. *Problem Statement:* A continuous source of contaminate-free water is needed.
 2. *Independent-Variable Conditions:*
    Energy
    Source
    Process
 3. *Methods of Satisfying Each Condition:*
    Types of Energy:
    a. Solar
    b. Electrical
    c. Fossil
    d. Atomic

  e. Mechanical
 Types of Source:
  a. Underground
  b. Atmosphere
  c. Surface supply
 Types of Process:
  a. Distillation
  b. Transport
  c. Manufacture
 4. *The Matrix* (Figure 11-3)
 5. *Combinations*

This particular matrix may be represented as an orderly arrangement of 45 small blocks stacked to form a rectangular parallelepiped. Every block will be labeled with the designations selected previously. Thus, block *X* in our preceding example suggests obtaining pure water by distilling a surface supply with a solar-energy power source, block *Y* means transporting water from an underground source by some mechanical means, and block *Z* recommends manufacturing water from the atmosphere using atomic power. Obviously, some of the blocks represent well-known solutions, and others suggest absurd or impractical possibilities. But some represent untried combinations that deserve investigation.

Where more than three variables are involved, electronic computers may be used to excellent advantage. After the matrix has been programmed, the computer can print a list of all the alternative combinations. Use of the computer is especially helpful when considering a large number of parameters.

The preceding techniques of stimulating new design concepts are particularly useful for the individual engineer. But often several designers may be searching jointly for imaginative ideas about some particular product. Then it is advantageous to use one of the procedures, "brainstorming" or "synectics."

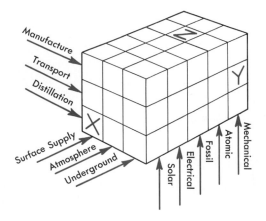

Figure 11-3

Brainstorming

The term "brainstorming" was coined by Alex F. Osborn[1] to describe an organized group effort aimed at solving a problem. The technique involves compiling all the ideas which the group can contribute and deferring judgment concerning their worth. This is accomplished (1) by releasing the imagination of the participants from restraints such as fear, conformity, and judgment; and (2) by providing a method to improve and combine ideas the moment an idea has been expressed. Osborn points out that this collaborative group effort does not replace individual ideative effort. Group brainstorming is used solely to supplement individual idea production and works very effectively for finding a large volume of alternative solutions or novel design approaches. It has been particularly useful for stimulating imaginative ideas for new products or product development. It is not recommended where the problem solution will depend primarily on judgment or where the problem is vast, complex, vague, or controversial. A homogeneous "status group" of six to twelve persons seems to be best for stimulating ideas with this method. However, the U.S. Armed Forces have used a hundred or more participants effectively. The typical brainstorming session has only two officials: a chairman and a recorder. The chairman's responsibility is to provide each panel member with a brief statement of the problem, preferably 24 hours prior to the meeting. He should make every effort to describe the problem in clear, concise terms. It should be *specific*, rather than *general*, in nature. Some examples of ideas that satisfy the problem statement may be included with the statement. Before beginning the session, the chairman should review the rules of brainstorming with the panel. These principles, although few, are very important and are summarized as follows:

1. **All ideas which come to mind are to be recorded.** No idea should be stifled. As Osborn says, "The wilder the idea, the better; it is easier to tame down than to think up." He recommends recording ideas on a chalkboard as they are suggested. Sometimes a tape recorder can be very valuable, especially when panel members suggest several different ideas in rapid succession.
2. **Suggested ideas must not be criticized or evaluated.** Judgments, whether adverse or laudatory, *must be withheld* until after the brainstorming session, because many ideas which are normally inhibited because of fear of ridicule and criticism are then brought out into the open. In many instances, ideas that would normally have been omitted turn out to be the best ideas.
3. **Combine, modify, alter, or add to ideas as they are suggested.** Participants should consciously attempt to improve on other people's ideas, as well as contributing their own imaginative ideas. Modifying a previously suggested idea will often lead to other entirely new ideas.

Those who dream by day are cognizant of many things which escape those who dream only by night.
—Edgar Allen Poe

"The horror of that moment," the king went on "I shall never, never forget!"
"You will, though," the queen said, "if you don't make a memorandum of it."
—Lewis Carroll

---

[1] Alex F. Osborn, *Applied Imagination* (New York: Scribner's, 1963), p. 151.

4. The group should be encouraged to think up a large quantity of ideas. Research at the State University of New York at Buffalo[2] seems to indicate that, when a brainstorming session produces more ideas, it will also produce higher-quality ideas.

The brainstorming chairman must always be alert to keep *evaluations* and *judgments* from creeping into the meeting. The spirit of enthusiasm that will permeate the group meeting is also very important to the success of the brainstorming session. The entire period should be conducted in a free and informal manner. It is most important to maintain, throughout the period, an environment where the group members are not afraid of seeming foolish. Both the speed of producing and recording ideas, and the number of ideas produced, help create this environment. Each panel member should bring to the meeting a list of new ideas that he has generated from the problem statement. These ideas help to get the session started. In general, the entire brainstorming period should not last more than 30 minutes to 1 hour.

The recorder keeps a stenographic account of all ideas presented and after the session lists them by type of solution without reference to their source. Team members may add ideas to the accumulated list for a 24-hour period. Later, the entire list of ideas should be rigorously evaluated, either by the original brainstorming group or, preferably, by a completely new team. Many of the ideas will be discarded quickly—others after some deliberation. Still others will likely show promise of success or at least suggest how the product can be improved.

Some specialists recommend that the brainstorming team include a few persons

Illustration 11-15
Not uncommonly, one team member's inspired idea will set off a chain reaction of ideas from other team members.

[2]Sidney J. Parnes and Arnold Meadow, "Effects of Brainstorming Instructions on Creative Problem Solving by Trained and Untrained Subjects," *Journal of Educational Psychology*, Vol. 50, No. 4 (1959), p. 176.

who are broadly educated and alert but who are amateurs in the particular topic to be discussed. Thus new points of view usually emerge for later consideration. Usually executives or other people mostly concerned with *evaluation* and *judgment* do not make good panel members. As suggested previously, particular care should be taken to confine the problem statement within a narrow or limited range to ensure that all team members direct their ideas toward a common target. Brainstorming is no substitute for applying the fundamental mathematical and physical principles the engineer has at his command. It should be recognized that the objective of brainstorming is to stimulate ideas—not to effect a complete solution for a given problem.

Dr. William J. J. Gordon[3] has described a somewhat similar method of group therapy for stimulating imaginative ideas, which he calls "synectics."

Synectics

This group effort is particularly useful to the engineer in eliciting a radically new idea or in improving products or developing new products. Unlike brainstorming, this technique does not aim at producing a large number of ideas. Rather, it attempts to bring about one or more solutions to a problem by drawing seemingly unrelated ideas together and forcing them to complement each other. The synectics participant tries to *imagine* himself as the "personality" of the inanimate object: "What would be my reaction *if I were that gear* (or drop of paint, or tank, or electron)?" Thus, familar objects take on strange appearances and actions, and strange concepts often become more comprehensible. A key part of this technique lies in the group leader's ability to make the team members "force-fit" or combine seemingly unrelated ideas into a new and useful solution. This is a difficult and time-consuming process. Synectics emphasizes the conscious, preconscious, and subconscious psychological states that are involved in all creative acts. In beginning, the group chairman leads the members to understand the problem and explore its *broad* aspects. For example, if a synectics group is seeking a better roofing material for traditional structures, the leader might begin a discussion on "coverings." He could also explore how the colors of coverings might enhance the overall efficiency (white in summer, black in winter). This might lead to a discussion of how colors are changed in nature. The group leader could then focus the group on more detailed discussion of how roofing materials could be made

---

The person who is capable of producing a large number of ideas per unit of time, other things being equal, has a greater chance of having significant ideas.
—J. P. Guilford

What good is electricity, Madam? What good is a baby?
—Michael Faraday

He that answereth a matter before he heareth it, it is folly and shame unto him.
—Proverbs xviii.13

No idea is so outlandish that it should not be considered with a searching but at the same time with a steady eye.
—Winston Churchill

---

[3]William J. J. Gordon, *Synectics* (New York: Harper & Row, 1961).

to change color automatically to correspond to different light intensities—like the biological action of a chameleon or a flounder. Similarly, the leader might approach the problem of devising a new type of can opener by first leading a group discussion of the word "opening," or he could begin considering a new type of lawn mower by first discussing the word "separation."

In general, synectics recommends viewing problems from various analogous situations. Paint that will not adhere to a surface might be viewed as analogous to water running off a duck's back. The earth's crust might be seen as analogous to the peel of an orange. The problem of enabling army tanks to cross a 40-ft-wide, bottomless crevass might be made analogous to the problem that two ants have in crossing chasms wider than their individual lengths.

Synectics has been used quite successfully in problem-solving situations in such diverse fields as military defense, the theater, manufacturing, public administration, and education. Where most members of the brainstorming team are very knowledgeable about the problem field, synectics frequently draws the team members from diverse fields of learning, so that the group spans many areas of knowledge. Philosophers, artists, psychologists, machinists, physicists, geologists, biologists, as well as engineers, might all serve equally well in a synectics group. Synectics assumes that someone who is imaginative but not experienced in that field may produce as many creative ideas as one who *is* experienced in that field. Unlike the expert, the novice can stretch his imagination. He approaches the problem with fewer preconceived ideas or theories, and he is thus freer from binding mental restrictions. (Obviously, this will not be true when the problem requires analysis or evaluation, where experience is a vital factor.) There is always present in the synectics conference an expert in the particular problem field. The expert can use his superior technical knowledge to give the team missing facts, or he may even assume the role of "devil's advocate," pointing out the weaknesses of an idea the group is considering. *All* synectics sessions are tape recorded for later review and to provide a permanent record.

Many believe that brainstorming comes to grips with the problem too abruptly while synectics delays too long. However, industry is using both methods successfully today.

Preparation of a model

Thus far in analyzing the design process we have considered identification of the problem, collection of information, and generation of ideas. The next phase is one of bringing together all the diverse parts (ideas, data, parameters, etc.) into a meaningful whole. As suggested on page 252, this procedure is called *synthesis,* and it is best known to the engineer as the process of modeling.

Psychologists and others who study the workings of the human mind tell us that we can think effectively only about simple problems and small "bits" of information. They tell us that those who master complicated problems do so by reducing them to a series of simple problems which can be solved and synthesized to a final solution. This technique consists of forming a mental picture of the entire problem, and then simplifying and altering this picture until it can be taken apart into manageable

Illustration 11-16

Synectics can be a source of innovative solutions.

components. These components must be simple and similar to concepts with which we are already familiar, to situations that we know. Such mental pictures are called *models*. They are simplified images of real things, or parts of real things—a special picture that permits us to relate it to something already known and to determine its behavior or suitability.

We are all familiar with models of sorts—with maps as models for a road system; with catalogs of merchandise as models of what is offered for sale. We have a model in our mind of the food we eat, the clothes we buy, and of the partner we want to marry.

We often form judgments and make decisions on the basis of the model, even though the model may not be entirely appropriate. Thus, the color of an apple may or may not be a sign of its ripeness, any more than the girl's apple-blossom cheeks and tip-tilted nose are the sign of a desirable girlfriend, (Illustration 11-17).

Engineering models are similar to sports diagrams that are composed of circles, squares, triangles, curved and straight lines, and other similar symbols which are used to represent a "play" in a football or basketball game (see Illustrations 11-20 and 11-21). Such geometrical models are limited because they are two dimensional and do not allow for the strengths, weaknesses, and imaginative decisions of the individual athletes. Their use, however, has proved to be quite valuable in simulating a brief action in the game and to suggest the best strategy for players should they find themselves in a similar situation.

An *idealized model* may emphasize the whole of the system and minimize its component parts, or it may be designed to represent only some particular part of the system. Its function is to make visualization, analysis, and testing more practical. The engineer must recognize that he is merely limiting the complexity of the problem in order to apply known principles. Often the model may deviate considerably from the true condition; and the engineer must, of necessity, select different models to represent the same real problem. Therefore, the engineer must view his answers with respect to the initial assumptions of the model. If the assumptions were in error, or if their importance was underestimated, then the engineer's analysis will not relate closely with the true conditions. The usefulness of the model to predict future actions must be verified by the engineer. This is accomplished by experimentation and testing. Refinement and verification by experimentation are continued until an acceptable model has been obtained.

Two characteristics, more than many others, determine an engineer's competence. The first is his ability to devise simple, meaningful models; and second is the breadth of his knowledge and experience with examples with which he can compare his models. The simpler his models are, and the more generally applicable, the easier it is to predict the behavior and compute the performance of the design. *Yet models have value only to the engineer who can analyze them*. The beauty and simplicity of a model of the atom, Figure 11-4 will appeal particularly to someone familiar with astronomy. A free-body diagram of a wheelbarrow handle (Figure 11-5) has meaning only to someone who knows how such a diagram can be used to find the strength needed in the handle.

Aside from models for "things," we can make models of situations, environments, and events. The football or baseball diagram is such a model. Another familiar model of this type is the weather map (Illustration 11-23), which depicts high- and low-pressure regions and other weather phenomena traveling across the country. Any meteorologist will tell you that the weather map is a very crude model for predicting weather, but that its simplicity makes the explanation of current weather trends

A damsel of high lineage and a brow
May-blossom, and a cheek of apple-blossom
Hawk-eyes; and lightly her slender nose
Tip-tilted like the petal of a flower.

TENNYSON

Illustration 11-17

Illustration 11-18
Some models are successful . . .

Illustration 11-19
. . . other models fail.

Illustration 11-20

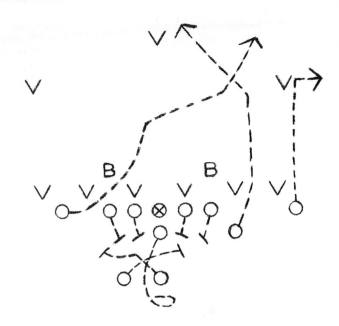

Illustration 11-21
Every football fan understands the value of diagrammed models in preparing for Saturday's "big game."

Illustration 11-22
Aeronautical engineers frequently work with models of their designs to confirm the validity of their calculations.

more understandable for the layman. Models of situations and environmental conditions are particularly important in the analysis of large systems because they aid in predicting and analyzing the performance of the system before its actual implementation. Such models have been prepared for economic, military, and political situations and their preparation and testing is a science all its own.

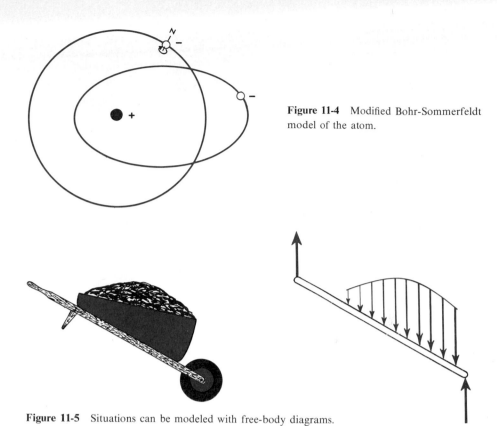

Figure 11-4 Modified Bohr-Sommerfeldt model of the atom.

Figure 11-5 Situations can be modeled with free-body diagrams.

Illustration 11-23
Weather maps are models of weather conditions.

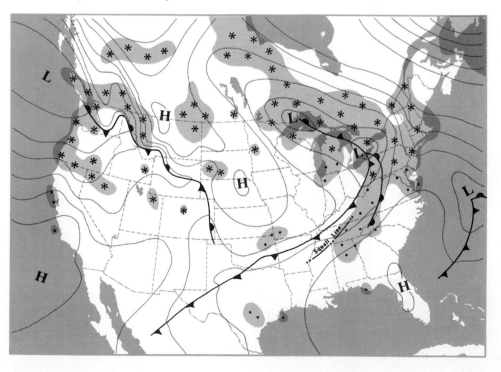

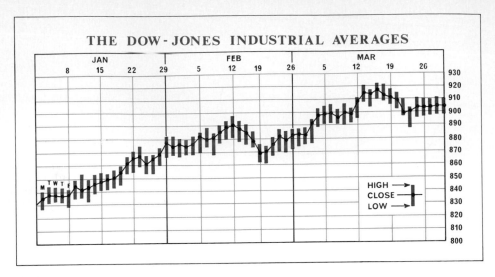

THE DOW-JONES INDUSTRIAL AVERAGES

Illustration 11-24
Many people utilize financial models.

Charts and graphs as models

Charts and graphs are convenient ways to illustrate the relationship among several variables. We have all seen charts of the fluctuations of the stock market averages (Illustration 11-24), in the newspaper from day to day, or you may have had your parents plot your growth on the closet door. In these examples, *time* is one of the variables. The others, in the examples above, are the average value of the stock in

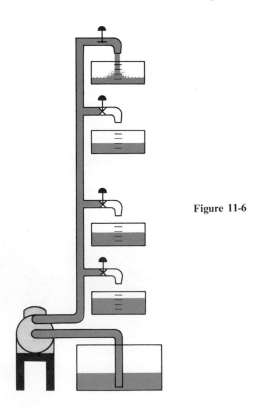

Figure 11-6

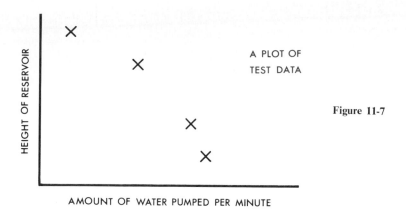

HEIGHT OF RESERVOIR

A PLOT OF
TEST DATA

Figure 11-7

AMOUNT OF WATER PUMPED PER MINUTE

dollars and your height in feet and inches, respectively. A chart or graph is not a model but presents facts in a readily understandable manner. *It becomes a model only when used to predict, project, or draw generalized conclusions* about a certain set of conditions. Consider the following example of how facts can be used to develop a chart and a graphical model. An engineer may wish to test a pump and determine how much water it can deliver to different heights (Figure 11-6). Using a stopwatch and calibrated reservoirs at different heights, the engineer measures the amount of water pumped to the different heights in a given time. The test results are plotted as crosses on a chart as shown in Figure 11-7. At this point, the plotted facts are a chart and not a model. Only when the engineer makes the assumption that the plotted points represent the typical performance of this or a similar pump under corresponding conditions can the chart be considered to be a graphical model. Once this assumption is made, the engineer can draw a smooth curve through the points. With this performance curve as a model, it is possible to predict that, if additional reservoirs are placed between the actual ones, they would produce results much like those shown by the circles in Figure 11-8. This assumption is made based on experience that pumps are likely to behave in a "regular" way. *Now* the graph is being used as an engineering model of the performance of the pump.

The diagram

A model often used by the engineer is the *diagram*. Typical forms of diagrams are the *block diagram,* the *electrical diagram,* and the *free-body diagram.* Some attention should be given to each of these forms.

The *block diagram* is a generalized approach for examining the whole problem, identifying its main components, and describing their relationships and interdependencies. This type of diagram is particularly useful in the early stages of design work when representation by mathematical equations would be difficult to accomplish. Figure 11-9 is an example of a block diagram in which components are drawn as blocks, and the connecting lines between blocks indicate the flow of information in the whole assembly. This type of presentation is widely used to lay out

> The battle of Waterloo was won on the playing fields of Eton.
> —Arthur Wellesley, Duke of Wellington

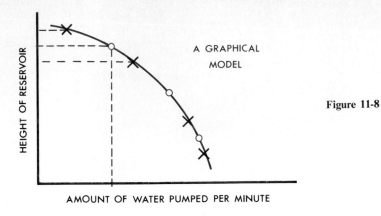

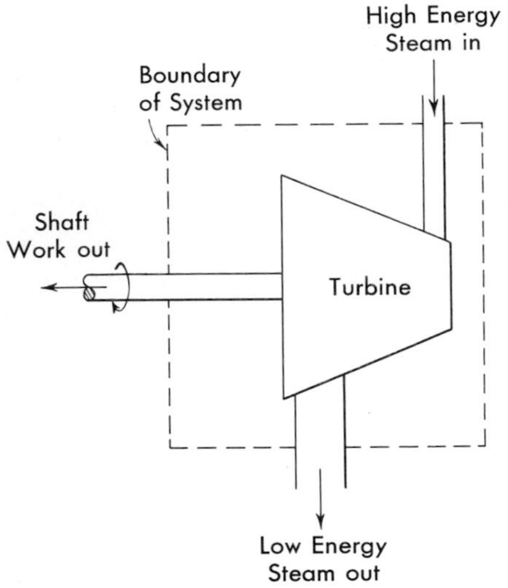

HEIGHT OF RESERVOIR
A GRAPHICAL MODEL
AMOUNT OF WATER PUMPED PER MINUTE

Figure 11-8

High Energy
Steam in
Boundary
of System
Shaft
Work out
Turbine
Low Energy
Steam out

Figure 11-9 Energy diagrams can be especially helpful in understanding thermodynamic systems.

Secondary Shield
Steam
Primary
Shield
Turbine
Reactor
Condenser
Water
Steam Generator
Primary
Coolant Pump
Feed Pump

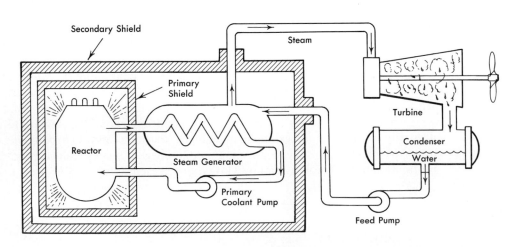

Figure 11-10 Diagrammatic sketch showing how nuclear power can be used to operate a submarine.

large or complicated systems—particularly those involving servoelectrical and mechanical devices. No attempt is made on the drawing to detail any of the components pictures. They are often referred to as "black boxes"—components whose *function* we know, but whose *details* are not yet designed.

The *energy diagram* is a special form of the block diagram and is used in the study of thermodynamic systems involving mass and energy flow. An example of the use of an *energy diagram*—is given in Figure 11-10.

The *electrical diagram* is a specialized type of model used in the analysis of electrical problems. This form of *idealized model* represents the existence of particular electrical circuits by utilizing conventional symbols for brevity. These diagrams may be of the most elementary type, or they may be highly complicated and require many hours of engineering time to prepare. In any case, however, they are representations or models in symbolic language of an electrical assembly.

Figure 11-11 shows an electrical diagram of a photoelectric tube that is arranged to operate a relay. Notice that the diagram details only the essential parts in order to provide for electrical continuity, and thus is an idealization that has been selected for purposes of simplification.

A *free-body diagram* (Figure 11-12) is a diagrammatic representation of a physical system which has been removed from all surrounding bodies or systems for purposes of examination and where the equivalent effect of the surrounding bodies is shown as acting on the free body. Such a diagram may be drawn to represent a complex system or any smaller part of it. This form of *idealized model* is most useful in showing the effect of forces that can act upon a system.

Illustration 11-25
The relation of component parts of a telemetering system are best shown by means of a block diagram.

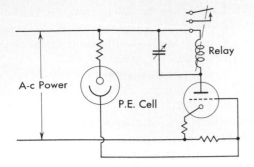

Figure 11-11   A simple photoelectric tube relay circuit.

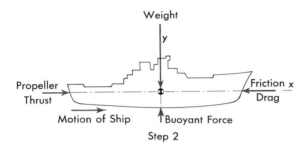

Figure 11-12   Free-body diagram of a ship.

The scale model

Scale models are used often in various problem-solving situations, especially when the system or product is very large and complex or very small and difficult to observe. A *scale model* is a replica, usually three dimensional, of the system, subsystem, or component being studied. It may be constructed to any desired scale relative to the actual design. Such projects as dam or reservoir construction, highway and freeway interchange design, factory layout, and aerodynamic investigations are particularly adaptable to study using scale models.

Scale models are useful for predicting performance because component parts of the model can be moved about to represent changing conditions within the system. Of considerably more usefulness are those scale models which are instrumented and subjected to environmental and load conditions that closely resemble reality. In such cases the models are tested and experimental data are recorded by an engineer. From an analysis of these data, predictions of the behavior of the real system can be made.

By using a scale model which can be constructed in a fraction of the time, a final design can be checked for accuracy prior to actual construction. Although scale models often cost many thousands of dollars, they are of relatively minor expense when compared with the total cost of a particular project.

Analog models

Analogs and similes are used to compare something that is unfamiliar to something else that is very familiar. Writers and teachers have found the *simile* to be a very effective way to describe an idea. Engineers use analogs in much the same way that teachers use similes. An analog, however, must provide more than a descriptive picture of what one wants to study; its action should correspond closely with the real

Illustration 11-26

thing. It should be *mathematically* similar to it, that is, the same type of mathematical expressions must describe well the action of both systems, the real and the analog.

A vibrating string is an analog of an organ pipe (Illustration 11-26). because the sound in an organ pipe behaves quite similarly to the waves traveling along a vibrating string. Under certain assumptions, similar mathematical equations can describe both systems. In other words, we can compare the corresponding actions of the *model* of the organ pipe with a *model* of the vibrating string. It is the *models* that behave exactly alike, *not the real systems*. If these models are "good" models, then, under certain conditions one can perform experiments with the string and draw valid conclusions concerning how the organ pipe would behave. Since one system may be much easier to experiment with than the other, one can work with the easier system and obtain results that are applicable to both.

An example of the use of a very successful analog is the electrical network that forms an analog for complete gas pipeline systems. Using such a model one can predict just what would happen if a lot of gas were suddenly needed at one point along the system. Experiments with the actual pipeline would be very costly and might disrupt service. The electrical network analog provides the answers faster, cheaper, and without disturbing anyone.

> Who so boasteth himself of a false gift is *an analog to* clouds and wind without rain.
> —Proverbs xxv:14

Analysis

One of the principal purposes of a model is to simplify the problem so that we can calculate the behavior, strength, and performance of the design. This is *analysis*.

Analysis is a mental process and, like any useful mental process, requires a store of basic knowledge and the ability to apply that knowledge. Since the amount of knowledge he possesses and his ability to use it are the major measures of a capable engineer, more time is spent at the university in studying analysis than any other subject. Just as one cannot solve a crossword puzzle without a knowledge of words, or make a medical diagnosis without a knowledge of the human body and its functions, so one cannot produce an acceptable engineering design without a basic knowledge of mathematics, physics, and chemistry, and their engineering relatives, such as stress analysis, heat transfer, electric network theory, and vibration. Nor can the engineer work effectively without an understanding of how his work affects man and his environment.

Analysis allows the engineer to "experiment on paper." For example, if he is concerned with the behavior of a wheel on a vehicle, his model might be a *rigid, perfectly round* wheel rolling on a *flat, unyielding* surface (Figure 11-13). (Words in italics indicate the assumptions made in the model.)

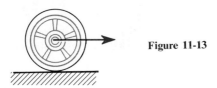

Figure 11-13

The motion of each point on the rim of the model wheel can be expressed by a well-known mathematical relationship called the *cycloid*. By knowing this motion the engineer can calculate the velocity of each point and determine how it varies with time (Figure 11-14). These calculations will enable him to solve for the centrifugal force on the wheel rim and to learn how fast the point makes contact with the ground.

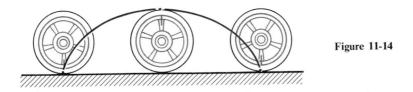

Figure 11-14

This elementary problem illustrates two important restrictions of engineering analysis: First, an equation usually describes only a very limited part of the function of the design, even in the case of a simple design such as a wheel. Second, an equation usually cannot describe *exactly* the action that takes place in the model. For example, we make the assumption that the wheel is perfectly rigid. This implies that it does not deform when it touches the ground. Such an assumption may be reasonably accurate in the case of a steel train wheel rolling on a steel rail, but it is probably a poor assumption to make for the rubber-tired wheels of an automobile.

Testing

The construction of a model and its analysis are based on assumptions, and these assumptions have to be verified.

One way in which the engineer can accomplish this is by the use of experiments. Experiments do not necessarily require construction of the entire design but only those portions of it that are important for the evaluation of the particular assumption. Components of a design are frequently tested instead of testing the entire design. If, in the design of a pipeline, the strength of the pipe is questionable, the engineer could obtain short sections of the pipe for testing. In the laboratory, liquid of appropriate density and pressure could be pumped through these sections of pipe to simulate the flowing fluid in the line. By measuring how much the pipe expands under the pressure and observing whether the results check with the calculations of his model, the engineer can verify if he has selected the proper pipe for his design.

Illustration 11-27

This rare 1859 engraver's print shows "Watt's First Experiment." As with engineers of today, young James Watt relied upon the results of experimentation to verify his theories. Here Watt is shown in the act of placing a spoon on the spout of a steaming teakettle to verify his observations that steam exerts pressure when confined. *One can imagine that the steam pushed young Watt's spoon away and it fell again, in rapid succession, to produce a trip-hammer sound. The artist has centered interest on the young engineer, who, neglectful of his lunch, intently watches the steam's effect. So also do his father and his aunt, Mrs. Muirhead, with a degree of kindliness which evidences regard for the boy, however much she may chide him for "listless idleness."*

It is important to remember that the value of the experiment is in the checking of the validity of the assumptions, not in checking the accuracy of the algebra. There is no need to run a rigid wheel on a rigid flat surface in order to prove the validity of the cycloidal motion of a point on the rim. If, however, the equation is to show the motion of a rubber-tired vehicle, then it may be well to run a rubber-tired wheel over a rigid surface to see how closely the cycloid does describe the motion of the nonrigid wheel.

For testing, one needs a model, a testing facility, and an arrangement of instruments suitable to measure what occurs during the test. Above all, one needs a test plan, just as a traveler needs directions in order to get to a desired destination. There is no sense in beginning a test without an objective and a plan for achieving the results necessary to satisfy that objective.

To test a completed design, the engineer should specify the characteristics that are most important and the instruments to be used for measuring these characteristics. In selecting the instruments the engineer must ask himself the question: "Does it provide the accuracy I need?" There being no such thing as *absolute* accuracy, the engineer must also know the probable error in the measurement. Only if that accuracy is greater than his allowable error is the instrument suitable. He is concerned also with the effect of the measurement on the performance itself. This is very important in the case of small, intricate devices requiring great accuracy. According to Heisenberg,[4] we cannot measure any characteristic without affecting the system. (We all know this to be true from our experience in a physician's office. Our pulse rate and blood pressure quite often change as soon as the doctor starts to measure them. Whether this is psychologically or physically conditioned does not really matter; the fact is the measurement *does* influence the performance.)

The conditions under which tests are to be conducted must be defined; these must include all the important conditions of the design. A list of characteristics that might be tested include startup and shutdown conditions, operation under partial and full load,[5] operation under the failure of auxiliary equipment, operator errors, material selection, and many, many others.

There are five frequently used objectives for engineering tests. These objectives determine: (1) quality assurance of materials and subassemblies, (2) performance, (3) life, endurance, and safety, (4) human acceptance, and (5) effects of the environment. Some tests are required on every design, and in certain cases all the tests are needed. In general, when the analysis has been completed a prototype model of the design will be constructed. Usually this model is subjected to all the necessary tests. Once the prototype has passed the tests and the design has gone into the production stage, each final product may be subjected to selected tests. In this case, the tests are primarily for the purpose of assuring product uniformity and reliability.

Prototype testing generally applies only to products which go into mass production, such as the automobile wheel. When the design is for a "one-of-a-kind" item such as a pipeline, one will make as many tests as possible on raw materials and subassemblies to detect design errors before construction is completed. However, final tests on the complete design still will be necessary *to ensure* its safety and acceptability.

[4]Werner Heisenberg (German physicist, 1901–   ) showed that, since observation must always necessarily affect the event being observed, this interference will lead to a fundamental limit on the accuracy of the observation. *Encyclopedia of Science* (New York: Harper & Row, 1967).

[5]Whenever we talk about *load,* we mean this in the general sense to include such things as force, pressure, voltage, vibration, amplitude, temperature, and corrosive effects.

Quality assurance tests

Anyone who has selected wood at a lumber yard knows that the quality of raw material varies substantially from one piece to the next. It is less well known that such variations occur also in other materials, such as metals, ceramics, and polymers. Variation in these materials may be as great as the variation between the pieces of wood at the lumber yard and, just as the lumberman will provide more uniform wood at a higher price, so one can get a more uniform steel, aluminum, or Plexiglas at a higher price (to pay for preselection by the manufacturer). Typical variations in the strength of metals can be found in engineering handbooks.

The competent designer should account for this type of variation—either by designing the part so that it will perform satisfactorily with the least desirable (weakest) material or by prescribing tests that would ensure that only premium materials be used. Both approaches add to the cost. The conservative design may require more material and more weight; the testing process may require the use of more expensive material, or the extra cost may result from the testing and the discarding of unusable pieces.

Manufacturers, like lumber dealers, have realized the need for uniformity in engineering materials. For this reason, materials with more uniform properties than standard, or with guaranteed minimum properties, are available (at higher prices). For example, one can buy electrical carbon resistors in three ranges: The first grade, indicated by a gold band, varies a maximum of 5 per cent from its indicated value; the second grade (silver band) may vary as much as 10 per cent; and the standard product (no band) may vary as much as 20 per cent. Typically a silver-band resistor will cost twice as much as the standard, and the gold four times as much. Quality assurance includes the checking of dimensions of completed parts (a type of inspection routine in most modern machine shops), tests on the quality of joints between two members whether welded, brazed, soldered, riveted, or glued, and the continuity of electrical circuits.

Performance tests

What has been said about raw materials is also true for components and subassemblies which the designer may wish to include in his design. Electric motors, pumps, amplifiers, heat exchangers, pressure vessels, and similar items are designed to certain manufacturing standards. The products usually will be constructed at least as well as the manufacturer claims. However, if the quality of the total design depends critically upon the specifications of a subassembly, then it is best to inspect and test that subassembly separately before it is included in the construction. This is particularly true for one-of-a-kind designs, such as space capsules. Since performance of the capsule is critically dependent upon that of its components, the designer must specify a series of tests that will be made at the manufacturers' plants to assure performance. He may, in fact, personally supervise the testing.

A performance test simply shows whether a design does what it is supposed to do. It measures the skill of the engineer and the validity of the assumptions made in his analyses.

Performance testing generally does not wait until the design is completed. It follows step by step with the design. For example, the heat shield of the space capsule is tested in the supersonic wind tunnel to see if it can withstand the aerodynamic

Illustration 11-28
All new products, such as this improved golf club, should be tested in use and their performance measured.

heating for which it was designed; the parachute is tested to see if it supports the capsule at just the right speed; structural members are tested for strength and stiffness; and instruments are checked to show if they indicate what they are supposed to measure.

Performance tests may require special testing apparatus, such as supersonic wind tunnels and space-simulation chambers. They always need careful planning and instrumentation to assure that the tests measure what is really needed—a proof of the validity of the design.

Life, endurance, and safety tests

We know that machines, like people, age. One of the most important and most difficult tests to gain meaningful data from are the life tests, tests that tell how long a product will survive in service and whether it can take excessive loads, misoperations, and other punishment without failure. It is rarely possible to carry out life tests accurately, for one seldom has the time to subject the prototype to the same period of aging that the real part will experience in actual service. In some instances, accelerated "life" tests are used. For example, paints and other surface protections may be exposed to the actions of sunlight, wind, rain, snow, or saltwater spray for months or even years. In this way a body of knowledge relating to the "life" of these surface finishes slowly develops, but a final selection by the engineer may be delayed considerably. Therefore, these tests must be accelerated and the engineer does so by increasing the load, by applying the load more rapidly, or by subjecting the design to a more severe environment. However, he is never quite sure how accurately these short-term "life" tests *really* represent the effect of the aging conditions and how their

Illustration 11-29
*Some tests need to be
continued until the
part actually fails.*

results should be interpreted. Usually, this is done (in the case of mechanical tests) by making tests on several specimens, each at a different degree of overload. The length of life of each part tested is then plotted as a function of the applied load, and the resulting curve is extrapolated to the maximum load that the part is expected to endure in real life.

If different types of loading are applied to the part, such as pressure, temperature, and vibration, it may be necessary to make separate tests with overloads in each one of these areas to see how they extrapolate to "true life." It may be desirable to use overloads in combination and observe if the combined effect is different from the sum of the effects of the individual loads. Very often two combined loads have a much more serious effect on the part than the arithmetic sum of the separate effects of the two loads. This is called *synergistic behavior.*

We again use the automobile wheel for an example of "life" testing. It is reasonable to assume that a good automobile tire and wheel should survive without failure at least 50,000 mi under normal loads, at speeds of 50 mi/hr. This means that the tire must be tested at that load for at least 1000 hr—a period in excess of 40 days. Such a test may indeed be possible and, in fact, is often performed on new tire designs. The designer will want to know the effects of overload, of speeds higher than 50 mi/hr, of curves, rough roads, under- and overinflation, and of the effect of very high or very low temperatures. It is easy to see that one needs a battery of testing machines to find out all one wants to know within a reasonable period of time.

Even though the test part may pass the predicted life during the test, the test is usually not terminated, but is continued until the part actually fails (Figure 11-15). This then becomes an endurance test, and it determines the excess life of the part. Since the life of each part is likely to differ, and since not every part can be "life" tested, it is essential to know the excess "life" of the average part. Since there is a statistical variability between parts, the engineer will want to know not only the *average* excess life but also the range of *variation* in this lifetime.

Instead of measuring the endurance of a part under a constant load, the engineer may decide to increase the load until the part fails. This failure load will be higher than the design load if the part is properly designed. The ratio between the failure

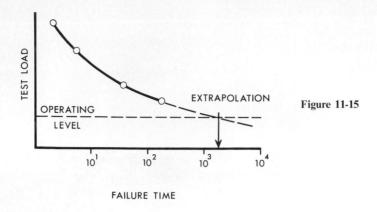

Figure 11-15

load and the design load is called the *factor of safety*. Factors of safety may be quite low where parts are very carefully manufactured, where excessive weight is undesirable, or where their failure causes no serious hardship. However, where human life is at stake, factors of safety must be so chosen that no variation in materials or workmanship, or simplified assumption in the designer's calculations, can possibly make the part unsafe and cause failure under normal operating loads.

Because the engineer has the responsibility to see that no one is injured as a consequence of his design, he must also consider the possibility of accidental or thoughtless misoperation of the design. For example, the automobile tire may be underinflated, it may be operated under too heavy a load, at too high a speed, or on a rough road, and yet it should not fail catastrophically.

Human acceptance tests

For a long time the designers of consumer goods have been concerned with the appearance and acceptability of their product to the buyer. It is unfortunate that in many instances they have appealed more often to the buyer's baser instincts, such as pride, greed, and desire for status, rather than to his sense of quality and beauty. Engineers, on the other hand, have been concerned too little with the interaction of their designs with the people who buy or use them. They have often failed to ask themselves whether the physical, mental, and emotional needs and limitations of the human being permit him to operate the machine in the *best* possible way. In the past engineers have all too often assumed that the human body is sufficiently adaptable to operate any kind of control lever or wheel. Although the adaptability of people is truly astonishing, we now know that for maximum efficiency levers and wheels must be carefully designed, that the forces necessary to operate them must be neither too large nor too small, that the operator should be able to "feel" the effect that he is producing, and that the use of the device or design should not tire him physically. A new design, then, should be tested with real people. Such tests usually cannot be performed by an engineer alone, but require the aid of others, such as industrial psychologists and/or human factors engineers.

Those who have studied metal processes have found that a man's attention span is short, that he cannot be asked continually to peer at a gage or work piece unless things are "happening" to it. It is also known that warnings are better heeded if they are audible than if they are visual. If they are visual, a bright red light or a blinking light which draws the operator's attention is much better than the movement of a dial to some position that has been marked "unacceptable." In recent years we have learned that our emotional responses play a major part in our daily lives, that whether we are elated or depressed can affect the quality of our work and the attention we give to the instruments we are asked to observe, or the levers we are asked to operate. In turn, our emotions are influenced by color, beauty, or attractive design—in short, by our opinion of the design.

The purpose of acceptability testing is to see whether the design meets the physical, mental, and emotional requirements of the average person for whom it is designed. Since the "average" person can never be found, it is essential that a series of tests be made by and with different people, and that these observations be used to make whatever changes are necessary to make the device as acceptable as possible within the technical and economic constraints.

Environmental tests

The environment is the aggregate of all conditions that surround the design under operating conditions. It may be wind and weather; it may be the soil in which the design is buried or the chemicals in which it is immersed; it may be the vacuum of outer space or an intense field of nuclear or electromagnetic radiation. In general, the operating environment is different from the normal environment in the laboratory in which the tests are made. Sometimes the effects of the environment are not important, but more frequently the environment can strongly affect the functioning and life of the part. Therefore environmental testing has become an important part of the final testing of most new products.

We have already mentioned how paints and other surface finishes are exposed to

Illustration 11-30
Some products, such as these engines, must operate under conditions of extreme heat and cold. Environmental testing is, therefore, mandatory.

sun, wind, and rain for long periods of time, thereby developing a considerable body of knowledge concerning how such surface finishes behave under both normal and unusual weather conditions. Similarly, an extensive body of knowledge exists on how chemicals deteriorate or "corrode" construction materials. These and other environmental factors are under continuing study. Therefore, unless the engineer is faced with an unusual environment, he can often (but not always) find pertinent information in the literature to predict how a particular environment is going to affect his design.

However, there are many instances where additional tests are needed. Tests are particularly important where two or more environmental effects work together, such as moisture and heat, or chemicals and vibrations. The result of such effects may not be predictable from either of the individual effects acting by themselves. Thus we know that a vibrating environment in a salt-spray atmosphere can cause corrosion fatigue at a rate far higher than that which might have been predicted from either the vibration or the saltwater corrosion taken independently.

With the advent of space travel, one of the most intensively studied types of environment is *space*. When away from the earth's atmosphere, a body in space will be in a nearly complete vacuum, but it will be exposed to a variety of types of radiation and to meteoric dust from which the earth's atmosphere normally protects it. The radiation effects may be severe enough to attack electronic circuits seriously and cause deterioration of transistors and other electronic devices. The meteoric dust, though generally quite fine, travels with speeds of 10,000 to 70,000 mi/hr and has sufficient energy to penetrate some of the strongest materials. Within the last few years some ways have been found to simulate in the laboratory both the high radiation and the presence of meteoric dust, and (on earth) to subject space equipment to these kinds of attack.

Safety and products liability

The engineer's role as a design specialist has always carried a sincere concern for the continued well-being of those who use his designs. In recent years the courts have taken a more positive role than heretofore in assessing the designer's responsibility concerning the health and safety of the user. Perhaps the most significant legislation pertaining to this matter to appear in modern times has been the Occupational Safety and Health Act (OSHA) which was signed into law on December 29, 1970.[6,7,8] This act sets forth the general conditions of safety which must be met for the manufacture or use of goods and services in this country. Basically it supplied the force of law to a host of previously defined "consensus standards" or codes that had been developed through the past 50 years by many professional, semi-professional, and governmental bodies. Although some of these codes, requirements, and recommended procedures were in keeping with modern engineering methods and materials, there were a host of others that were incorporated that were archaic and (although *technically* the latest printed standard available) were in need of modernization and

---

[6]R. J. Redding, *Intrinsic Safety* (New York: McGraw-Hill, 1976).
[7]Dan Petersen, *The OSHA Compliance Manual* (New York: McGraw-Hill, 1976).
[8]Peter S. Hopf, *Designer's Guide to OSHA* (New York: McGraw-Hill, 1976).

change. Nevertheless the engineer-designer must exercise great care to be certain that his designs meet the legal requirements that pertain to the product or system that he has designed.

Products liability is civil liability to an ultimate consumer for injury to his person or property resulting from using a defective article that has been sold by a person in the business, or chain, of selling such articles.[9]

In the Middle Ages, the rule of "buyer beware" worked satisfactorily for all concerned because each manufacturer's products were used primarily in the local community. Thus, a swordsmaker, silversmith, or bootsmith was generally identified by the *quality* of his handicraft. Transactions were normally consummated in face-to-face encounters. This all changed with the advent of the industrial age, and today products made in a small factory in California may be used extensively in every part of the country. Also, the identity of the manufacturer and/or designers may be unknown.

For this reason those who engage in a particular field of manufacturing are held responsible for possessing the knowledge and skill of an expert in that field and they must keep reasonably abreast of methods and procedures used by practical men in the trade. This responsibility includes due care to conduct reasonable tests and to discover latent hazards.

It is incumbent upon the manufacturer to warn potential users or consumers when it is known that the use of a particular product in a certain way would be hazardous. The engineering field is complex and changes in available technical capability and materials selection occur rapidly. Because of this, design parameters may be adjudged as safe today, but unsafe tomorrow. The engineer must do his best to foresee conditions of possible failure or hazard. It is also very important that his records be accurate and complete, and that they reflect that he has designed the product to be as safe as possible, as the existing state of the art and his imagination will allow.[10]

[9]Wyatt Jacobs, "Products Liability," *Mechanical Engineering*. November 1972, p. 12.
[10]*Encyclopaedia of Occupational Health and Safety*, Vols. I and II (New York: McGraw-Hill, 1976).

> If you make people think they're thinking, they'll love you; but if you *really* make them think, they'll hate you.
> —Don Marquis

Problems

11-1. After reviewing the ecological needs of your home town, state three problems that should be solved.

11-2. Give three examples of *feedback* that existed prior to 1800.

11-3. Talk to an engineer who is working in design or development in industry. Describe two situations in his work where he has not been able to rely on *theoretical textbook solutions* to solve his problems. Why was he forced to resort to other means to solve the problems?

11-4. Describe an incident where an individual or group abandoned their course of action because it was found that they were spending time working on the wrong problem.

11-5. List the properties of a kitchen electric mixer.

11-6. List the properties of the automobile that you would like to own.

11-7. Make a matrix analysis of the possible solutions to the problem of removing dirt from clothes.

11-8. For ten minutes solo brainstorm the problem of disposal of home wastepaper. List your ideas for solution.

11-9. List five types of models that are routinely used by the average American citizen.

11-10. Diagram the model of the football play that made the longest yardage gain for your team this year.

11-11. Draw an energy system of an ordinary gas-fired hot-water heater.

11-12. Draw an energy system representing a simple refrigeration cycle.

11-13. Draw an energy system representing a "perpetual motion" machine.

11-14. Draw an electrical circuit diagram containing two single-pole, double-throw switches in such a manner that a single light bulb may be turned on or off at either switch location.

11-15. Arrange three single-pole, single-throw switches in an electrical circuit containing three light bulbs in such a manner that one switch will turn on one of the bulbs, another switch will turn on two of the bulbs, and the third switch will turn on all three bulbs.

11-16. Describe three situations where a scale model would be the most appropriate kind of idealized model to use.

11-17. How can engineering help solve some of the major world problems?

11-18. Discuss some of the inventions that have contributed to the success of man's first lunar exploration.

11-19. Write a paragraph entitled "Fiction Today, Engineering Tomorrow."

11-20. Propose a method and describe the general features of a value system whereby we could replace the use of money.

11-21. List five problems that might now confront the city officials of your home town. Propose at least three solutions for each of these problems.

11-22. Cut out five humorous cartoons from magazines. Recaption each cartoon such that the story told is completely changed. Attach a typed copy of your own caption underneath the original caption for each cartoon.

11-23. Propose a title and theme for five new television programs.

11-24. The following series of five words are related such that each word has a meaningful association with the word adjacent to it. Supply the missing words.

Example	girl	*blond*	*hair*	*oil*	rich
a. astronaut		_____	_____	_____	engineer
b. pollution		_____	_____	_____	automobile
c. college		_____	_____	_____	textbook
d. football		_____	_____	_____	radio
e. food		_____	_____	_____	energy

11-25. Suggest several "highly desirable" alterations that would encourage personal travel by rail.

11-26. What are five ways in which you might accumulate a crowd of 100 people at the corner of Main Street and Central at 6 A.M. on Saturday?

11-27. "As inevitable as night after day"—using the word "inevitable," contrive six similar figures of speech. "As inevitable as . . . ."

11-28. Name five waste products, and suggest ways in which these products may be reclaimed for useful purposes.

11-29. Recall the last time that you lost your temper. Describe those things accomplished and those things lost by this display of emotion. Develop a strategy to regain that which was lost.

11-30. You have just been named president of the college or university that you now attend. List your first ten official actions.

11-31. Describe the best original idea that you have ever had. Why has it (not) been adopted?

11-32. Discuss an idea that has been accepted within the past ten years but which originally was ridiculed.

11-33. Describe some design that you believe defies improvement.

11-34. Describe how one of the following might be used to start a fire: (*a*) scout knife, (*b*) baseball, (*c*) pocket watch, (*d*) turnip, (*e*) light bulb.

11-35. At night you can hear a mouse gnawing wood inside your bedroom wall. Noise does not seem to encourage him to leave. Describe how you will get rid of him.

11-36. Write a jingle using each of these words: cow, scholar, lass, nimble.

11-37. You are interviewing young engineering graduates to work on a project under your direction. What three questions would you ask each one in order to evaluate his creative ability?

11-38. Describe the most annoying habit of your girlfriend (boyfriend). Suggest three ways in which you might tactfully get this person to alter that habit for the better.

11-39. Suggest five designs that are direct results of ideas that have been stimulated by each of the five senses.

11-40. "A man's mother is his misfortune; his wife is his own fault." *The London Spectator.* Write three similar epigrams on boy–girl relations.

11-41. Put a blob of ink on a piece of paper and quickly press another piece of paper against it. Allow it to dry, and then write a paragraph describing "what you see in the resulting smear."

11-42. List the criteria for an urban transportation system in your town. Assign relative values to each of the criteria.

11-43. It has been found that for comfort, train passengers should not be exposed to accelerations or decelerations greater than about 4.5 feet/second^2 in the direction of motion. For urban rapid transit trains, which must stop every 2 miles, this restriction (and not top speed) sets the limit on how long it takes to get from station to station. So says Mr. L. K. Edwards, president of Tube-Transit, Inc. He proposes instead to dig inclined tunnels and allow gravity to speed up (and slow down) the trains. He says he can achieve much greater real accelerations without the passengers feeling any acceleration at all. He proposes that the slope be about 15° (Figure 11-16).

Evaluate the merits of his scheme and find the time it would take between stations 2 miles apart for Edward's train as against a regular, aboveground train.

11-44. Make a checklist and attribute list for each of the following:

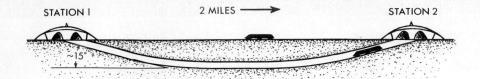

STATION 1　　　　　2 MILES ⟶　　　　　STATION 2

Figure 11-16

      (a) house telephone   (d) shower faucet
      (b) ironing board   (e) can opener
      (c) light switch

11-45. Design a new mechanism to deploy the Lunar antenna shown in Illustration 11-31.

11-46. Make a literature search and prepare a written and oral report on the following subjects:
      (a) Packaging of a TV tube for shipment.
      (b) Transmission of electric power to a high-speed train
      (c) Automatic bicycle transmissions
      (d) A speed reduction device (gearbox) with at least 10:1 reduction
      (e) Nonslip highway surfacing
      (f) Solar-powered refrigeration

11-47. Prepare a two- or three-dimensional matrix of the independent conditions for the problems in 11-46 and show how at least six of the possible combinations might be used.

Illustration 11-31

11-48. (Note to the instructor: The following problems are related. They are intentionally vague and ill defined like most real-life problems. Their purpose is to stimulate creativity and imaginative solutions, to permit students to find out for themselves, make assumptions, test them, compare ideas, build models, and prepare reports— written or oral—to convince a nontechnical audience. Give only as much aid or additional information as you believe to be absolutely essential. Additional problems for this setting may suggest themselves.)

You are a Peace-Corps volunteer (or a small team) about to be sent to a village of about 500 people in a primitive, underdeveloped country. The village lies 3,000 feet below a steep escarpment in a valley through which a raging river flows. The river is about 80 feet wide, 4 to 8 feet deep, and too fast to wade or swim across. On your side of the river there is the village of mud huts in a clearing of the hardwood forest. The trees are no more than 40 feet tall. At the foot of the escarpment there is broken rock. Across the river there is another village, which cannot be reached except by a very long path and a difficult river crossing upstream. There are other villages on top of the escarpment. The people are small, few over 5 feet 6 inches tall. They live mostly by hunting, gathering, and fishing, though they could trade to their benefit with the people across the river and on the escarpment if communication were easier.

Before you leave for your assignment, you should try to find solutions to one or more of the following problems:

(a) How to improve communication, trade, and social contact between the two villages on each side of the river.

(b) How to transport goods easily up and down the escarpment. There is a path up the escarpment, but it is steep, dangerous, and almost useless as a trade route.

(c) Suggest a better way of hunting than with the bow and arrows now used. A crossbow has been suggested to be more powerful, easier to aim, and more accurate. Evaluate these claims and provide design criteria.

(d) Provide for lighting of the huts. The villagers now use wicks dipped in open bowls of tallow. Can you improve their lamps so that they burn brighter, smoke less, and don't get blown out in the wind?

For each of these problems, select criteria for evaluating ideas. Choose several different solutions, check them against the criteria, and pick the best one; develop this idea by analysis and testing until you know how it will work. All the while, keep track of and test your assumptions whenever possible. Finally, prepare a way to convince the villagers of the value of your idea.

12

Engineering design phases

Much of the history of man has been influenced by developments in engineering, science, and technology. When progress in these fields was impeded, the culture of the era tended to stagnate and decline; the converse was also true. Although many definitions have been given of "engineering," it is generally agreed that *the basic purpose of the engineering profession is to develop technical devices, services, and systems for the use and benefit of man.* The engineer's design is, in a sense, a bridge across the unknown between the resources available and the needs of mankind (Figure 12-1).

Regardless of his field of specialization or the complexity of the problem, the method by which the engineer does his work is known as the *engineering design process.* This process is a creative and iterative approach to problem solving. It is creative because it brings into being new ideas and combinations of ideas that did not exist before. It is iterative because it brings into play the cyclic process of problem solving, applied over and over again as the scope of a problem becomes more completely defined and better understood.

> . . . the process of design, the process of inventing physical things which display new physical order, organization, form, in response to function.
> —Christopher Alexander, *Notes on the Synthesis of Form*
>
> A scientist can discover a new star but he cannot make one. He would have to ask an engineer to do it for him.
> —Gordon L. Glegg

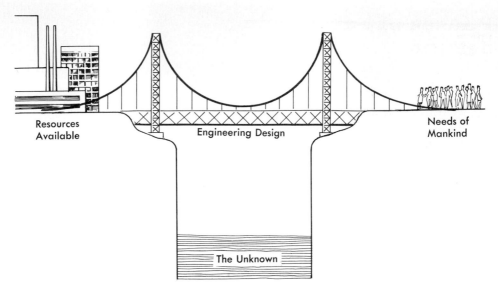

Resources Available

Engineering Design

Needs of Mankind

The Unknown

Figure 12-1

Thus a design engineer must be a creative person—an idea man—and he must be able to try one idea after another without becoming discouraged. In general he learns more from his failures than from his successes, and his final designs usually will be compromises and departures from the "ideal" that he would like to achieve.

A final engineering design usually is the product of the inspired and organized efforts of more than one person. The personalities of good designers vary, but certain characteristics are strikingly similar. Among these will be the following:

1. Technical competence.
2. Understanding of nature.
3. Empathy for the requirements of his fellowmen.
4. Active curiosity.
5. Ability to observe with discernment.
6. Initiative.
7. Motivation to design for the pleasure of accomplishment.
8. Confidence.
9. Integrity.
10. Willingness to take a calculated risk and to assume responsibility.
11. Capacity to synthesize.
12. Persistence and sense of purpose.

Certain design precepts and methods can be learned by study, but the ability to design cannot be gained solely by reading or studying. The engineer also must grapple with real problems and apply his knowledge and abilities to finding solutions. Just as an athlete needs rigorous practice, so an engineer needs practice on design problems as he attempts to gain proficiency in his art. Such experience must necessarily be gained over a period of years, but now is a good time to begin acquiring some of the requisite fundamentals.

Phases of engineering design

Most engineering designs go through three distinct phases:

1. The feasibility study.
2. The preliminary design.
3. The detail design.

In general, a design project will proceed through the various phases in the sequence indicated (Figure 12-2). The amount of time spent on any phase is a function of the complexity of the problem and the restrictions placed upon the engineer—time, money, or performance characteristics.

The feasibility study

The feasibility study is concerned with the following:

1. Definition of the elements of the problem.
2. Identification of the factors that limit the scope of the design.
3. Evaluation of the difficulties that can be anticipated as probable in the design process.
4. A realistic appraisal of the return (profit) on investment.
5. Consideration of the consequences of the design.

The objectives of the feasibility study are to discover possible solutions and to determine which of these appear to have promise and which are not feasible, and why.

Let us see how this might work in a situation where you, as the chief engineer of an aircraft company, have been asked to diversify the company's product line by designing a small, low-power passenger vehicle for town driving with substantially less pollution than present cars.

What are the elements of the problem; what factors limit its scope? Where and by whom will such vehicles be used? Are they to carry people individually and randomly to and from work, school, shopping areas, or places of amusement like the present car, or are they to be links in a more comprehensive transportation system? Are they to be privately or publicly owned? If privately owned, perhaps the emphasis should be on low cost, simple upkeep, and ease of parking. If publicly owned—a car which a licensed driver can pick up at one parking place and leave at another—then ease of

Figure 12-2

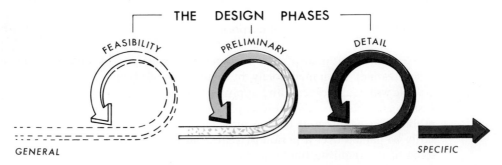

THE DESIGN PHASES

FEASIBILITY   PRELIMINARY   DETAIL

GENERAL   SPECIFIC

handling, reliable operation, and long life might be the major considerations. You will want to know how fast it is to go, how far between fuelings, and how many people it is to carry.

Assume that it is decided to design a vehicle for public ownership. What difficulties can be anticipated? Probably the major ones will have to do with people and what they might do. How do you make such a vehicle safe and nearly foolproof? People must be prevented from driving it too far and from abandoning it anywhere except at designated parking places. Provision must be made to redistribute the vehicles if for some reason—a ball game, a sale, a happening—too many people converge on one area. Maintenance and repair will present many problems.

Assuming that such a vehicle can be built and sold to cities, what would be the consequences? Some of the desirable ones are obvious: less traffic, lower air pollution, fewer parking problems, and more efficient vehicle utilization. But what about the uncertainty of finding a car when and where you want it, particularly on a wet, cold night or during the rush hour? The new rules and regulations that would have to be devised? The risk of nonacceptance by the public? These would be some of the early considerations during such a feasibility study.

The ideas and possibilities which are generated in early discussions should be checked for the following:

1. Acceptability in meeting the specifications.
2. Compatibility with known principles of science and engineering.
3. Compatibility with the environment.
4. Compatibility of the properties of the design with other parts of the system.
5. Comparison of the design with other known solutions to the problem.

Each alternative is examined to determine whether or not it can be physically achieved, whether its potential usefulness is commensurate with the costs of making it available, and whether the return on the investment warrants its implementation. The feasibility study is in effect a "pilot" effort whose primary purpose is to seek information pertinent to all possible solutions to the problem. After the information has been collected and evaluated, and after the undesirable design possibilities have been discarded, the engineer still may have several alternatives to consider—all of which may be acceptable.

During the generation of ideas, the engineer has intentionally avoided making any final selection so as to have an open mind for all possibilities and to give free rein to the thought processes. Now the number of ideas must be reduced to a few—those most likely to be successful, those that will compete for the final solution. The number of ideas kept will depend on the complexity of ideas and the amount of time and manpower that he can afford to spend during the preliminary design phase. In most design situations the number of ideas remaining at the end of the feasibility study will vary from two to six.

At this point no objective evaluations are available; the discarding of ideas must

> The successful producer of an article sells it for more than it cost him to make, and that's profit. But the customer buys it only because it is worth *more* to him than he pays for it, and that's his profit. No one can long make a profit *producing* anything unless the customer makes a profit *using* it.
> —Samuel B. Pettengill

depend to a large extent upon experience and judgment. There are few substitutes for experience, but there are ways in which judgment can be improved. For example, decision processes based on the theory of probability can be employed effectively. Analog and digital simulations are particularly useful to the engineer in this early comparison of alternatives.

In some instances, it will be more convenient for the engineer to compare the expected performance of the component parts of one design with the counterpart performances of another design. When this is done, it is very important to consider if the component parts create the optimum effect in the overall design. Frequently it is true that a simple combination of seemingly ideal parts will not produce an optimum condition. It is not too difficult to list the advantages and disadvantages of each alternative, but the proper evaluation of such lists may require the wisdom of Solomon. Economic feasibility is also a requirement of all successful designs.

What is economics?

Economics is that social science that is concerned primarily with the description and analysis of the problems of production, distribution, and use of goods and services. In the United States today, products are not sold because they have been made; they are made because they have been sold. And, they are sold because there is a demand for them. Demand means that at a specific level of product price, a certain number of units can be marketed. The number that can be marketed will determine (in general) the production facilities and processes that are needed.

The economic problem

Any economic problem always has two aspects—production and distribution. Every society has had to fashion some kind of system to *produce* the goods and services that its members need or want. Also every society has had to fashion some kind of system to *distribute* the goods and services that are produced. At different times, and in different places, the ways in which a particular society has solved its economic problem have varied sharply.

Solutions to the economic problem

Historically, there have been three solutions or methods of controlling the market:

1. Tradition.
2. Command.
3. Market enterprise (sometimes called *free enterprise*).

Often these solutions do not have sharp boundaries. Each type exists somewhere in the world today, and aspects of each of them can be observed in every society.

> There can be no economy where there is no efficiency.
> —Benjamin Disraeli, 1804–1881

Traditional economies. In traditional economies each family unit provides for most of its own needs. The hunters and gatherers of earliest human history lived from the wild fruits and grains that they could gather and the animals that they could kill. In later more advanced agrarian societies, families learned to plant food and raise domestic animals. In both types of societies there was little trade, and that which did exist was barter. For example, one family with extra corn might trade it to a neighboring family for milk. In this way production was controlled by the needs of each family and the extent to which it was willing to work in order to fill or exceed these needs.

Command societies. The early command societies were based on the existence of conscripted labor or slavery. Slaves were obtained as a by-product of conquest, by piracy and kidnapping, or, later, by commerce. They were employed as domestics, public servants, artisans, musicians, and teachers, as well as on the farms, in the mines, and in commerce and manufacturing. At the time of the emperor Claudius it was estimated that there were nearly 21 million slaves in all of Italy. The point is that, except for the small allowances paid them, this was all *free* labor.

In modern times in command economies the society (government) itself owns the tools of production and prescribes what is to be produced, how much, and at what price it is to be sold. Russia and China use this method of controlling the market. The important quality to note about both traditional and command societies is that in each case, *wealth follows power,* whereas in market-enterprise societies *power follows wealth.*

Market-enterprise economies. In market-enterprise societies the quality that regulates what is to be produced, and how much, is *price*. This price is the price at the point of final use. In other words, every time a housewife buys a can of peas and the transaction is recorded by the supermarket cashier, she is voting on whether another

Illustration 12-1
*In a market
enterprise society,
customers young
and old daily
exercise judgments
affecting the
success or failure
of the product.*

similar can of peas should be produced (Illustration 12-1). The government did not "order" the production of any peas, let alone that specific brand. The producer continues to exist solely because enough housewives vote "yes" in this way. In most cases, they register their favorable votes on the question at hand primarily on the basis of their own family's satisfaction with the product. What is true of the can of peas is equally true of home appliances, lathes, generators, automobiles, and virtually every other product currently available in the American market. Just how all of this affects the engineer and the design process will become clearer as we proceed.

Hallmarks of market economies

The United States market represents the most highly developed form of market-enterprise society in the world today. It is a *mass* market of over 230 million people in 50 states between which there are few real restraints on commerce. Consequently, it is relatively easy for anyone with an idea to organize an enterprise to produce and distribute a new product. Although sometimes it seems that this is a country of big business and big unions, it should be noted that currently there are over 300,000 manufacturing enterprises that have fewer than 200 employees each. Over 20,000 new incorporations are recorded each month by the Office of Business Economics, U.S. Department of Commerce. The same agency reports about 35 failures each month for every 10,000 firms that are engaged in business (incorporated and unincorporated). Most fail because not enough attention has been given in the planning phase to the stark realities of market-enterprise economics. We shall discuss below some of the reasons for business successes and failures.

The United States is a society of *contract* rather than a society of *status*. In the time of Thomas Jefferson the laws of *primogeniture* were repealed. Primogeniture was a principle, inherited from the British, in which the eldest son inherited the father's estate. By repeal of this principle we have established the principle that young people should have the right to define their life stations for themselves, rather than having their stations defined for them by means of inheritance. However, in this

> Which of you, intending to build a tower, sitteth not down first, and counteth the cost, whether he have sufficient to finish it?
> —Luke 14:28, *The Holy Bible*
>
> Money never starts an idea; it is the idea that starts the money.
> —W. J. Cameron
>
> Of all human powers operating on the affairs of mankind, none is greater than that of competition.
> —Senator Henry Clay, *Address before the U.S. Senate,* February 2, 1832

system the same right that is possessed by one person is also possessed by his competitors. Every American is free to seek gain as he or she wishes, rather than being "locked in" a particular status by reason of his birth. John Locke[1] argued that "every man has a property in his own person. The labor of his body, and the work of his hands ... are properly his." Similarly, Adam Smith[2] declared that "the property which every man has in his own labor (as it is the original foundation of all other property) so it is the most sacred and inviolable."

A market-enterprise society cannot exist where there is slavery. If, for example, one fourth of the society kept the other three fourths enslaved, then three fourths of that society would have virtually no income. Under such conditions a market-enterprise economy cannot develop, since the only buyers available to purchase goods are the one fourth of the society that has some expendable revenue.

In a market-enterprise society nearly everyone who works has some income. Therefore, some monetary reward is associated with almost all tasks. The whole society provides a potential market for most of the products that are produced. This condition results in the basic principle of market-society economics: *Quantity of the product demanded is a function of price.* This principle is the implicit regulator that controls the economic constraints in a market-enterprise society.

Where does economics enter the design process?

We often hear engineers talk about "R and D," which is a shorthand designation for research and development. Generally, the *research* phase of the design process is considered to be complete upon the development of the *prototype*—the experimental model from which the final design is developed. This early model should be studied carefully to make certain that the design can be manufactured, and this process is what is meant by *development*. It is at this point that certain economic considerations become critical to the future success of the enterprise. The most pertinent economic considerations are those concerned with the realities of the "market," which regulates the production and distribution of goods.

Market-enterprise parameters for the designer

While engaged in the actual physical activity of creating a design, the designer must

---

[1] John Locke, "Of Property," Chapter V of *An Essay Concerning the True Original Extent and End of Civil Government,* 1690.
[2] Adam Smith, *Wealth of Nations,* Book I (New York: Modern Library, 1937), p. 121.

keep in mind the following constraints:

1. *The total market.*
2. *That portion of the total market which might "demand" a product like the one that he is designing.*
3. *That portion of the reduced market which might demand his particular design.* This is called "market penetration." As an example, the Kirsch Company of Sturgis, Michigan, not long ago sold 65 per cent of all the venetian blind hardware that was marketed in the United States. In this case its market penetration was 65 per cent.
4. *The price at which competitive products are being sold.* The price at which one's design can be sold is not merely the manufacturing cost plus some expected rate of return. Rather, the *price of competitive products* usually will determine the maximum allowable cost to make and market a new design.
5. *The basic price/sales relation for the product.* For example, more automobiles can be sold for $8000 each than for $15,000 each.

Production and marketing

Thus far we have discussed the economic problem, the different types of economic systems, and especially the market-enterprise solution to the problem.

What are the chances for success of a new product?

New products are the lifeblood of commerce. What can be said about the chances of success of a new product? Looking at the whole picture, 98 per cent of all new products introduced to the general market fail within two years. There is an inexhaustible list of reasons for this. Poor design, poor packaging, poor market research, inexperienced management, insufficient capital, lackadaisical selling effort, and failure to provide maintenance facilities are some of the more prominent reasons.

Among companies that are experienced in the design and introduction of new products to the market, about one new product out of five proves successful. But in many cases one out of five is enough to ensure fame and fortune for the innovators and to provide the capital that market enterprise needs to grow on.

These statistics should encourage the young engineer to consider every aspect of design, packaging, and marketing strategy *before* committing individual or company capital.

What are the chances for success of a new company?

In recent years more than 400,000 new firms are started each year. However, of the vast total of United States firms more than 350,000 are being discontinued annually, and ownership or control is being transferred in a slightly larger number. The relative frequency of outright failure, however, varies greatly between types of business.

Dun & Bradstreet, which keeps a record of such facts, recently asked, "What are the primary causes of business failure?" Here are the answers they found.

	Per cent
Incompetence	45.6
Unbalanced experience	19.5
Lack of managerial experience	13.7
Lack of experience in the line	8.7
Neglect	2.8
Disaster	1.4
Fraud	1.2
Reason unknown	7.1
Total	100.0

Unbalanced experience means that the firm's experience was not well rounded in sales, finance, purchasing, and production on the part of the management unit. Thus nearly 90 per cent of the failures were caused by the management's incompetence. And, of these failures, over half occurred in firms five years old or younger.

It should be recognized that the above failures occurred during one of the longest sustained boom periods in our history. For business there has never been a period quite like that of the years 1960–1980. And yet even during this period of expansion and high volume in almost every line of business, something over 40 per cent of all the manufacturing firms in operation showed either no profit or an actual operating loss.

There are as many business objectives as there are human motives, but the primary objective of any operating enterprise must be *profit*. We have said that in a market-enterprise economy the individual must be left free to pursue gain. This is just as true for the firm as it is for the individual. No matter what the other objectives of a company might be, nothing can be accomplished unless that company first makes certain that it earns a profit on its operations. A "normal" rate of profit must be considered by the designer as an element of the total cost. The economist considers profit to be a residue after expense is subtracted from revenue. And the accountant, who has the responsibility of determining income for tax purposes, is not likely to count profits until they have already occurred. Nevertheless, the designer must plan for profits before the fact, as a part of the design process.

The consumer is a capricious taskmaster. There is no explaining the fickle nature of customer taste, and often the best effort of the designer falls before a subtle change in fashion. There is no known way in which the designer can ensure, beforehand, that the effort devoted to design will gain that magical response called *consumer acceptance*. However, the root of company success is product success. Even today, with its hundreds of thousands of employees and with its thousands of separate products, the General Motors Corporation views its basic product as the internal

Illustration 12-2

The Mustang—a marketing success.

combustion engine. Whatever else this company does, it makes very, very sure of the
excellence of design and performance of this basic device.

The life cycle of new products

The life span of a new product depends upon the type of product and may vary
from a few years to many decades. In Congressional hearings on the drug industry a
few years ago, one of the major pharmaceutical manufacturers testified that 95 per
cent of all the products in its current catalog were less than five years old. Because of
such competitive conditions within the pharmaceutical industry, there is tremendous
pressure to develop and market new products at a very high rate. However, this also
means a very rapid rate of obsolescence of the products that are currently in production. At the other end of the spectrum we have what are referred to as *producers'
goods,* such as water turbines for power generation, or the heavy steel rolls for billet
mills in the steel industry. Tools such as these are more likely to be objects of continual improvement, rather than of overnight obsolescence.

The life of most successful products follows a trend similar to that shown in
Figure 12-3. This type of *growth curve* is seen often. It describes the typical growth of
populations—of people, plants, and animals. Frequently, it also describes new-prod-
uct life cycles, such as those of the steamboat, the steam locomotive, the automobile,
and television. Such a curve has four basic sections. In the early stages it rises at a
slow but increasing rate. Then there follows a period during which production is
proportional to the passage of time. We usually call this section the *linear* portion of
the curve. Then there follows a section late in time, when production is still increasing, but at a decreasing rate. Finally, as with the buggy and then with the steam
locomotive, production declines.

What can the designer do to prevent this decay? As time passes and production
increases, continuous improvement should be made in the design. New materials
should be tested, new technical improvements sought, and perhaps new packaging
and marketing strategies investigated. By such modifications the product's usefulness

Illustration 12-3
The Edsel—a marketing failure.

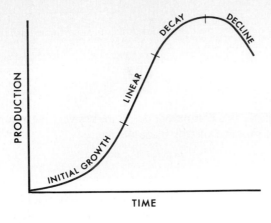

Figure 12-3 Typical growth history.

may be extended to new operating environments or perhaps totally new uses can be found for it in its present environment. The addition of color has greatly extended the linear portion of the TV life cycle. Combination of TV with computers and with telephones may extend that life even further.

After the designer has completed the product prototype, and after a manufacturable version of it has been developed, it is time to turn to improvement, simplification, and cost reduction. In these ways the onset of *incipient decline* can be markedly delayed. And this is one reason that design is such an exciting activity—it is always immersed in action.

Optimization

The demands of production and of marketing—as well as the correction of the mistakes described—all have two factors in common: they all require substantial allocation of capital, and they all require time. Both of these factors are what the economists call *scarce resources*. Some companies spend most of their capital on engineering and enter bankruptcy after designing the finest product of its kind. Other companies choose to emphasize production to the exclusion of both engineering and marketing. Other firms are extremely sales minded and tend to sacrifice engineering and production on the altar of increased volume. Lack of proper balance in any of these aspects can be disastrous.

So we see that there are conflicting demands for capital. When we have resolved these conflicts into the best possible overall results under the circumstances, we say that we have *optimized* the performance. This requires what are called trade-offs among the competing demands for a limited supply of capital. If one spends availa-

> Industry prospers when it offers people articles which they want more than they want anything they now have. The fact is that people never buy what they need. They buy what they want.
> —Charles F. Kettering

Illustration 12-4
An understanding of the value and use of money is important to the engineer. Usually the amount of money available for a design is less than that desired to accomplish the task to be performed.

ble funds on new machinery, he may not have what he would like to spend on advertising. If we spend our money on inventory to be certain that the factory never runs out of supplies, we may not have enough money to meet the payroll.

How to tell how you are doing

We shall discuss only two of the myriad tools used by the operating businessman to determine whether he is optimizing his overall performance. Many professional and trade associations maintain statistical services for their members, but the two tools we shall mention are absolutely fundamental and therefore are generally available.

Fourteen important ratios. Every year in the November issue of *Dun's Review and Modern Industry,* Dun & Bradstreet publishes fourteen important ratios for many lines of manufactures. This company for many, many years has devoted itself to collecting and maintaining files on American businesses. Each year those firms interested in getting and keeping a good credit rating submit their accounting reports, both balance sheets and operating statements, to Dun & Bradstreet. In addition to assigning a credit rating to the respondent firm, Dun & Bradstreet tabulates the data by industry for the benefit of all its subscribers. In recent years the tabulation has been by SIC code number, making it still more useful. SIC is shorthand for Standard Industrial Classification, a publication of the government.

To find out if you are optimizing your overall performance (as compared with

> When a man says money can do anything, that settles it: he hasn't any.
> —Ed. Howe
>
> Time is money.
> —Benjamin Franklin
>
> Money is a stupid measurement of achievement but unfortunately it is the only universal measure we have.
> —Charles P. Steinmetz

your competitors) all you have to do is to look up the SIC code number for your major product in the SIC manual. Then you can turn to the Dun & Bradstreet compilation of the fourteen ratios and find your particular industry. Under this entry you will find the best, median, and poorest performances for your industry. You will find how your industry performed in ratio of current assets to current debt, sales to inventory, profit to sales, profit to net worth, sales to net worth, and nine other ratios that have come to be regarded as valuable indicators of business performance. Based upon the same ratios from your own accounting reports, you can take corrective action.

Census of manufactures. In every year ending in a 2 or a 7, the U.S. Department of Commerce conducts a census of manufactures. It is such a huge job that it takes a year or two to tabulate the data and to begin to issue the findings. For this reason these data are not as current as those of Dun & Bradstreet, but they are just as useful, nonetheless. The census lists data for each of 20 industries, for geographical areas right down to the level of the county in many cases, and for many specific products. Of particular value are the data on value of shipments, value added by manufacture, number of production workers, and man-hours and earnings of production workers. When we divide value added through manufacture by production workers, production worker man-hours, or production worker earnings, we obtain three basic measures of productivity to guide us in our design and development activities. As before, the need for corrective action may be indicated by a comparison of our own firm's data with that for the industry.

By carefully using the data published by Dun & Bradstreet and the Census of Manufactures, the designer can get some idea of how new product ideas may fare in the market. At the very least, it is possible to identify those new product ideas which are likely to fail.

A tentative selling price

The calculation of a tentative price for a proposed design should be given consideration as a part of the feasibility design study. The primary objective of arriving at a tentative selling price so early in the design process is to make certain that it is possible to achieve the recovery of all costs plus some "specified" return on the company's investment. In general, the costs of production will be a function of the overall investments made in buildings and machinery, the skill classes of labor used, and the processes required in the production of the design. Usually, a number of different manufacturing processes can be employed to achieve the same design result. For instance, a machine part might be manufactured by forging, stamping, casting, or machining. Although a finished part might perform the same function without regard to its method of manufacture, the respective cost to achieve each of the various processes could vary considerably. Also, the initial investment in equipment necessary to accomplish each of the various processes might vary widely, and so might the costs of operation of the equipment. A company's equipment investment will depend upon the complexity of the design and upon the anticipated quantity of the product. Usually, choices will have to be made between securing special or general-purpose machinery, but it may be that the company already has available a sufficient amount of equipment that can be used to produce the design without making any new investment at all.

If new equipment must be purchased, and if a high production rate is anticipated, it might be advantageous to modify the design—for example, so that the part could be stamped from relatively cheap sheet metal instead of being cast or machined. For stamping, the special dies required by a punch press, although expensive initially, may more than pay their way in reducing the labor costs per piece by making possible high production rates. On the other hand, one may wish to utilize presently available special-purpose machines, such as automatic lathes, in which case allowance must be made for the increased material and labor costs. Or the company may have a foundry that could take advantage of the relatively cheap labor and material costs and produce the part by casting.

The design engineer should make decisions only after considering the overall results expected by his or her management. Such results will be a function of the competitive patterns of the industry in which one works. In determining a final selling price, it is not realistic to simply add up a design's detailed costs, add the desired profit, and announce, "The price shall be thus-and-so." In the final analysis a producer's selling price must take into account the competitor's selling price, if competition does exist. This price must also provide a realistic margin on a product's sales price and one that will yield a specified return on the investment. The specified return on investment is the basic guide for optimization of the overall performance of a firm.

Most men believe that it would benefit them if they could get a little from those who *have* more. How much more would it benefit them if they would learn a little from those who *know* more.
—W. J. H. Boetcker

Profits decline more during (an economic) crisis than wages. It is because of the fall in profits that unemployment occurs.
—Jean Fourastié

Value

The consideration of *value* is very important in the early selection process. From whose point of view should a particular alternative be appraised? Performance characteristics that may be advantageous in one situation may be equally disadvantageous in another. As an example, automatic redistribution of cars would increase the efficiency of the public car system and save driver cost. However, such an automatic system would almost surely not be possible on public streets, and the cost of extra rights-of-way may make it prohibitive. How does one select the location and proximity of parking places? How far should people be asked to walk, and how many parking places can be serviced effectively? How does one select the maximum speed of the cars and reconcile the conflicting demands of safety and service? Where danger to human life is a possibility, the measurement of value becomes exceedingly difficult. There is great reluctance to place a "cost" or value on the life of a human being. If the engineer assumes an infinite cost penalty, the design may be impossible, but to ignore this factor would effectively assign a cost factor of zero to a life. The engineer must face every responsibility with honesty and realism.[3]

Engineers engaged in a feasibility study must be able to project the future effectiveness of the alternative designs. In many cases the preliminary design stage of a product will precede its manufacture by several years. Conditions change with time, and these changes must be anticipated by the engineer. Many companies have become eminently successful because of the accuracy of their projections, whereas others have been forced into bankruptcy.

The preliminary design phase

With alternatives narrowed to a few, the engineer must select the design he wishes to develop in detail. The choice is easy if only one of the proposed designs fulfills all requirements. More often, several of the concepts appear to meet the specifications equally well. The choice then must be made on such factors as economics, novelty, reliability, and the number and severity of unsolved problems.

Since it is difficult to make such comparisons in one's head without introducing personal bias, it is useful to prepare an evaluation table. All the important design criteria are listed, and each is assigned an importance factor. There always will be both positive and negative criteria. Then each design is rated as to how well it meets each criterion. This rating should be done by somebody who is not aware of the value assigned to each importance factor, so that he is not unduly influenced.

Let us apply this procedure to our city transportation problem, and particularly to the selection of the propulsion system. Let us assume that the ordinary automobile engine has already been discarded because it is unable to meet air pollution requirements, and that the choice has narrowed to one of three types of engines: the gas turbine, the electric motor, and the steam engine. We will then enter these as Designs (1), (2), and (3) in a table and assign values to the various positive and negative design criteria (Table 12-1). For example, the gas turbine and electric motor rate low on "novelty" for they are well developed, but an automobile steam engine could rate

---

[3] By assigning financial damages to families whose breadwinner has lost his life in an industrial accident, the courts have effectively placed a monetary value on human life. Damages as high as $250,000 have been awarded.

Table 12-1 *Evaluation of propulsion systems. Importance (I) varies from 1 (small importance) to 5 (extreme importance). Rating (R) values are 3 (high), 2 (medium), 1 (low), and 0 (none).*

Design criteria	Importance I	Design (1) gas turbine		Design (2) electric		Design (3) steam	
		R	R × I	R	R × I	R	R × I
Positive							
a. Novelty		0		1		3	
b. Practicability		1		3		2	
c. Reliability		2		3		1	
d. Life expectancy		2		2		2	
e. Probability of meeting specifications		2		3		2	
*f. Adaptability to company expertise (research, sales, etc.)		1		1		1	
*g. Suitability to human use							
*h. Other							
Total positive score		—	—	—	—	—	—
Negative							
a. Number and severity of unresolved problems		1		2		3	
b. Production cost		3		1		2	
c. Maintenance cost		1		1		2	
d. Time to perfect		1		1		3	
* Environmental effects		1		0		1	
* Other							
Total negative score		—		—		—	
Net score		—		—		—	

* Such factors may not always be pertinent.

high if it used modern thermodynamic principles. On "practicability" the electric motor rates higher than the others, for it requires the least service and provides the easiest and safest way to power a small vehicle. This table is completed to the best ability of the engineer for each of the criteria.

Then the engineer "blanks out" the ratings and assigns "importance" factors to each of the criteria (Table 12-2). For example, it may be appropriate to rate practicability much higher than novelty.

Finally the ratings and importance factors are multiplied and added, yielding a final rating for the three systems (Table 12-3), which, in this case, favors the electric motor drive. Although others may come up with different ratings, the method minimizes personal bias.

After selecting the best alternative to pursue, the engineer should make every effort to refine the chosen concept into its most elementary form. Simplicity in design has long been recognized as a hallmark of quality. Simple solutions are the most difficult to achieve, but the engineer should work to this end. It is important to remember that such timeless ideas as the lever, the wedge, the inclined plane, the screw, the pulley, and the wheel are still basic ingredients of good design.

In terms of the electric drive vehicle, this means that initially he will strive for a single motor, directly driving the rear wheels, and a battery that can be recharged in

Table 12-2   Evaluation of propulsion systems. Importance (I) varies from 1 (small importance) to 5 (extreme importance). Rating (R) values are 3 (high), 2 (medium), 1 (low), and 0 (none).

Design criteria	Importance I
Positive	
a. Novelty	2
b. Practicability	5
c. Reliability	5
d. Life expectancy	3
e. Probability of meeting specifications	4
*f. Adaptability to company expertise (research, sales, etc.)	3
*g. Suitability to human use	N.A.
*h. Other	
Total positive score	
Negative	
a. Number and severity of unresolved problems	3
b. Production cost	4
c. Maintenance cost	4
d. Time to perfect	4
* Environmental effects	4
* Other	
Total negative score	
Net score	

* Such factors may not always be pertinent.

each parking area. He may later find that a smaller motor at each wheel is preferable, that a geared-down, high-speed motor is more efficient than a direct-drive motor, or that an on-board electric generator is preferable to a rechargeable battery. He will start with the simplest ideas.

Once the design concept has been selected, the engineer must consider all the component parts—their sizes, relationships, and materials. In selecting materials, strengths, dimensions, and the loads to which they will be exposed must be considered. In this sense, the engineer is analogous to the painter who has just chosen his subject and now must select his colors, shapes, and brush strokes and put them together in a pleasing and harmonious arrangement. The engineer, having selected a design concept that fulfills the desired functions, must organize his or her components to produce a device that is not only pleasing to the eye but is economical to build and operate.

Engineers must make sure that their designs do not interfere with or disturb the environment, that it agrees with man and nature. We are especially reminded of these responsibilities when we encounter foul air, polluted streams, and eroded watersheds. Environmental effects are increasingly important criteria in the design of engineering structures, as evidenced by the voluble concern about such projects as the transAlaska pipeline, the supersonic jet transport, and facilities for the disposal

Table 12-3 *Evaluation of propulsion systems. Importance (I) varies from 1 (small importance) to 5 (extreme importance). Rating (R) values are 3 (high), 2 (medium), 1 (low), and 0 (none).*

Design criteria	Importance I	Design (1) gas turbine		Design (2) electric		Design (3) steam	
		R	R × I	R	R × I	R	R × I
Positive							
a. Novelty	2	0	0	1	2	3	6
b. Practicability	5	1	5	3	15	2	10
c. Reliability	5	2	10	3	15	1	5
d. Life expectancy	3	2	6	2	6	2	6
e. Probability of meeting specifications	4	2	8	3	12	2	8
*f. Adaptability to company expertise (research, sales, etc.)	3	1	3	1	3	1	3
*g. Suitability to human use	N.A.						
*h. Other							
Total positive score			32		53		38
Negative							
a. Number and severity of unresolved problems	3	1	3	2	6	3	9
b. Production cost	4	3	12	1	4	2	8
c. Maintenance cost	4	1	4	1	4	2	8
d. Time to perfect	4	1	4	1	4	3	12
* Environmental effects	4	1	4	0	0	1	4
* Other							
Total negative score			27		18		41
Net score			5		35		−3

° Such factors may not always be pertinent.

or reclamation of industrial and human waste. As the earth's natural resources are depleted, the engineer will be under increasing pressure to provide technical assurances that no harm is done to the environment.

The designer must consider such factors as heat, noise, light, vibration, acceleration, air supply, and humidity, and their effects upon the physical and mental well-being of the user. For example, while it would be desirable to accelerate to top speed as quickly as possible, there are human comfort limits on acceleration that should not be exceeded. Controls must respond rapidly, have the right "feel," and not tire the driver. The suspension system must be "soft" for a comfortable ride, but stiff enough for good performance on curves. Automatic heating and air conditioning will probably be required in most parts of the country.

By now the picture of the vehicle has become clearer, and the chief engineer can delegate the preliminary design of components to various engineers or designers in the organization. Someone will be working on the drive train, another on the wheels and suspension, a third on the battery. Then there are the speed control systems, the interior layout, and perhaps three or four other components, such as access protection, recharging, and systems for redistributing the cars that must be developed.

The detail design phase

Detailed design begins after determination of the overall functions and dimensions of the major members, the forces and allowable deflections of load-carrying members, the speed and power requirements of rotating parts, the pressures and flow rates of moving fluids, the aesthetic proportions, and the needs of the operation—in short, after the principal requirements are determined. The models that were devised during the preliminary selection process should be refined and studied under a considerably wider range of parameters than was possible originally. The designer is interested not only in normal operation, but also in what happens during startup and shutdown, during malfunctions, and in emergencies. The range of the loads which act on a design and how these loads are transmitted through its parts as stresses and strains must be evaluated. The effects of temperature, wind, and weather, of vibrations and chemical attack should be considered. In short, the range of operating conditions for each component of the design and for the entire device must be determined.

Design engineers must have an understanding of the mechanisms of engineering: the levers, linkages, and screw threads that transfer and transform linear and rotating motion; the shafts, gears, belts, and chain drives that transmit power; the electrical power generating systems and their electronic control circuits.

With today's wide range of available materials, shapes, and manufacturing techniques, with the growing array of prefabricated devices and parts, the choices for the design engineer are vast indeed. How should you start? What guidelines are available if you want to produce the best possible design? It is usually wise to begin investigating that part or component which is thought to be most critical in the overall design—perhaps the one that must withstand the greatest variation of loads or other environmental influences, the one that is likely to be most expensive to make, or the most critical in operation. You may find that operating conditions limit your choices to a few possibilities.

At this stage as the designer you will encounter many conflicting requirements. One consideration tells you that you need more power, another that the motor must be smaller and lighter. Springs should be stiff to minimize road clearance; they should be soft to give a comfortable ride. Windows should be large for good visibility, but small for safety and high body strength. The way to resolve this type of conflict is called optimization. It is accomplished by assigning values to all requirements and selecting that design which maximizes (optimizes) the total value.

Materials and stock subassemblies are commercially available in a specific range of sizes. Sheet steel is commonly available in certain thicknesses (gages), electric motors in certain horsepower ratings, and pipe in a limited range of diameters and wall thicknesses. Generally, the engineer should specify commonly available items; only rarely will the design justify the cost of a "special mill run" with off-standard dimensions or specifications. When available sizes are substantially different from the desired optimum size, the engineer may have to revise his optimization procedure.

To illustrate, let us look at the design of a meteorological rocket. At an earlier point in the design process the fuel for this rocket will have been chosen. Let us assume that it is a solid fuel, a material that looks and feels like rubber, burns without air, and when ignited produces high-temperature, high-pressure gases which are

---

A civilization is both developed and limited by the materials at its disposal . . . .
—Sir George Paget Thomson

expelled through the nozzle to propel the rocket. The rocket consists principally of the payload (the meteorological instruments that are to be carried aloft), the nose cone which houses the instruments, the fuel, the fuel casing, and the nozzle. If we can estimate the weight of the rocket and how high it is to ascend, then we can calculate the requirements.

The most critical design part is the fuel casing, that is, the cylindrical shell which must contain the rocket fuel while it burns. It must be strong enough to withstand the pressure and temperature of the burning fuel, and strong enough to transmit the thrust from the nozzle to the nose cone without buckling and without vibrating. The shell must also be light. If the casing weighs more than had been estimated originally, then more fuel will be needed to propel the rocket. More fuel will produce higher pressures and higher temperatures inside the casing. This, in turn, will require a stronger casing and even more weight. This additional weight requires still more fuel, and the spiral continues.

Let us assume we decided to use a high-strength, high-temperature-resistant steel for our casing. Our calculations indicate its wall thickness to be not less than 0.28 in. Our steel catalog tells us this steel is generally available in sheet form only in thicknesses of $\frac{1}{4}$ and $\frac{3}{16}$ in. If we use the thicker sheet, the casing weight will increase by 2.7 per cent; then we must recalculate the amount of fuel required, the pressures and stresses in the casing, and consequent changes in the dimensions of the rocket. Will the $\frac{1}{4}$-in. material withstand the resultant higher stresses? Can we improve its strength by heat treating? If we choose the thinner material, must we provide the casing with extra stiffeners (rings which will reduce the stresses in the casing shell)? In either case, the original design must be altered until the stresses, weights, pressures, and dimensions are satisfactory.

Similar design procedures will be followed in designing the nose cone, the nozzle, and the launching gear for the rocket.

It is important to understand that this example is typical of the design process. Design is not a simple straightforward process but a procedure of *trial and error and compromise* until a well-matched combination of components has been found. The more the engineer knows about materials and about ways of reducing or redistributing stresses (in short, the more alternatives that are available the better the structural design is likely to be).

Consider, as another example, that as the engineer you have been asked to design the gear shift lever for a racing automobile. The gear box has already been designed, so you know how far the shifting fork (the end that actually moves the gears in the gear box) must travel in all directions. You also know how much force will be required at the fork under normal and abnormal driving conditions. You will need to refer to anthropometric[4] data to learn how much force the healthy driver can provide forward, backward, and sideways, and what his reach can be without distracting his eye from the road. With all this information you can choose the location of the ball joint, the fulcrum of the gear shift lever, and the length of each arm of the lever. You may decide to use a straight stick or you may find that a bent lever is more convenient for the driver. Before you finalize this decision you may build a mock-up and make experiments to determine the most convenient location. Next you must select the material and the cross-sectional shape and area of the lever. Since it is likely to be loaded evenly in all directions, you may find that a circular or a cruciform cross

---

[4] *Anthropometry* is the study of human body measurements, especially on a comparative basis.

section is most suitable. You must decide between a lever of constant thickness and a lighter, tapered stick (with the greater strength where it is needed—near the joint) which is more costly to manufacture.

Next you will consider the design of the ball joint, which transmits the motion smoothly to the gear box and provides vibration isolation so that the hand of the driver does not shake. It is difficult to find just the right amount of isolation which will retain for the driver the "feel" that is so essential during a race. Finally, you will need a complete understanding of lubricated ball joints and proficiency in testing a series of possible designs.

The final component in this design is the handle itself, which should be attractive to look at and comfortable to grip. Here again anthropometric data can tell you much, yet you will be well advised to make several mock-ups and to have them tested for "feel" by experienced drivers.

During the design process, you will have made a series of sketches (somewhat like those in this book) to illustrate to yourself the relative position of the parts that you are designing. Now you or your drafter will use these sketches to make a finished drawing. This will consist of a separate detail drawing for each individually machined item, showing all dimensions, the material from which it is to be made, the type of work to be performed, and the finish to be provided. There also will be subassembly and assembly drawings showing how these parts are to be put together.

The detail design phase will include the completion of an operating physical model or prototype (a model having the correct layout and physical appearance but constructed by custom techniques), which may have been started in an earlier design phase. The first prototype usually will be incomplete and modifications and alterations will be necessary. This is to be expected. Problems previously unanticipated may be identified, undesirable characteristics may be eliminated, and performance under design conditions may be observed for the first time. This part of the design process is always a time of excitement for everyone, especially the engineer.

The final phase of design involves the checking of every detail, every component, and every subsystem. All must be compatible. Much testing may be necessary to prove theoretical calculations or to discover unsuspected consequences. Assumptions made in the earlier design phases should be reexamined and viewed with suspicion.

Illustration 12-5
You oaf! You misread the scale again. I wanted a toy for my son. Now what could we ever do with a wooden horse this big?

Are they still valid? Would other assumptions now be more realistic? If so, what changes would be called for in the design?

As one moves through the design phases—from feasibility study to detail design—the tasks to be accomplished become less and less abstract and consequently more closely defined as to their expected functions. (See Illustration 12-2.) In the earlier phases, the engineer worked with the design of systems, subsystems, and components. In the detail design phase he also will work with the design of the parts and elementary pieces that will be assembled to form the components.

In the previous phase of engineering design, a large majority of the people involved were engineers. In the detail phase this is not necessarily the case. Many people—metallurgists, chemists, tool designers, detailers, draftsmen, technicians, checkers, estimators, manufacturing and shop personnel—will work together under the direction of engineers. These technically trained support people probably will outnumber the engineers. The engineer who works in this phase of design must be a good manager in addition to his technical responsibilities, and his successes may be measured largely by his ability to bring forth the best efforts of many people.

The engineer should strive to produce a design which is the "obvious" answer to everyone who sees it, *once it is complete*. Such designs, simple and pleasing in appearance, are in a sense as beautiful as any painting, piece of sculpture, or poem, and they are frequently considerably more useful to his well-being.

The planning of engineering projects

In every walk of life, we notice and appreciate evidence of well-planned activities. You may have noticed that good planning involves more than "the assignment of tasks to be performed" although this frequently is the only aspect of planning that is given any attention. Planning in the broad sense must include the enumeration of all the activities and events associated with a project and a recognition and evaluation of their interrelationships and interdependencies. The assignment of tasks to be performed and other aspects of scheduling should follow.

Since "time is money," planning is a very important part of the implementation of any engineering design. Good planning is often the difference between success and failure, and the young engineering student would do well, therefore, to learn some of the fundamental aspects of planning as applied to the implementation of engineering projects.

In 1957 the U.S. Navy was attempting to complete the Polaris Missile System in record time. The estimated time for completion seemed unreasonably long. Through the efforts of an operations research team, a new method of planning and coordinating the many complex parts of the project was finally developed. The overall saving in time for the project amounted to more than 18 months. Since that time a large percentage of engineering projects, particularly those which are complex and time consuming, have used this same planning technique to excellent advantage. It is called PERT (Program Evaluation and Review Technique).

PERT enables the engineer in charge to view the total project as well as to recognize the interrelationships of the component parts of the design. Its utility is not limited to the beginning of the project but rather it continues to provide an accurate

> Though this be madness, yet there is method in it.
> —Shakespeare

measure of progress throughout the work period. Pertinent features of PERT are combined in the following discussion.

How does PERT work?

Basically PERT consists of events (or jobs) and activities arranged into a *time-oriented network* to show the interrelationships and interdependencies that exist. One of the primary objectives of such a network is to identify where bottlenecks may occur that would slow down the process. Once such bottlenecks have been identified, then extra resources such as time and effort can be applied at the appropriate places to make certain that the entire process will not be slowed. The network is also used to portray the events as they occur in the process of accomplishing missions or objectives, together with the activities that necessarily occur to interconnect the events. These relationships will be discussed more fully below.

The network A PERT network is one type of pictorial representation of a project. This network establishes the "precedent relationships" that exist within a project. That is, it identifies those activities which must be completed before other activities are started. It also specifies the time that it takes to complete these activities. This is accomplished by using *events* (points in time) to separate the project *activities*. In other words, project events are connected by activities to form a project network. Progress from one event to another is made by completing the activity which connects them. Let us examine each component of the network in more detail.

Events An event is the *start* or *completion* of a mental or physical task. It does not involve the actual performance of the task. Thus, events are *points in time* which require that action be taken or that decisions be made. Various symbols are used in industry to designate events, such as circles, squares, ellipses, or rectangles. In this book circles, called *nodes,* will be used (Figure 12-4).

Events are joined together to form a project network. It is important that the events be arranged within the network in logical or time sequence from left to right. If this is done, the completion of each event will occupy a discrete and identifiable point in time. An event cannot consume time and it cannot be considered to be completed until all activities leading to it have been completed. After all events have been identified and arranged within the network, they are assigned identification numbers. Since events and activities may be altered during the course of the project, the logical order of the events will not necessarily follow in exact numerical sequence, 1, 2, 3, 4, 5, and so on. The event numbers, therefore, serve only for identification purposes.

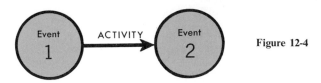

Figure 12-4

The final or terminal node in the network is usually called the *sink,* while the beginning or initial node is called the *source.* Networks may have varying numbers of sources and sinks.

Activities An activity is the actual performance of a task and, as such, it consumes an increment of time. Activities separate events. An activity cannot begin until all preceding activities have been completed. An arrow is used to represent the time span of an activity, with time flowing from the tail to the point of the arrow (Figure 12-5). In

Activity   **Figure 12-5**

a PERT network an activity may indicate the use of time, manpower, materials, facilities, space, or other resources. A *phantom* activity also may represent waiting time or "interdependencies." A phantom activity, represented by a dashed arrow, (Figure 12-6), may be inserted into the network for clarity of the logic, although it

Phantom Activity   **Figure 12-6**

represents no real physical activity. Waiting time would also be noted in this manner. Remember that:

> *Events* "happen or occur."
> *Activities* are "started or completed."

The case of Mr. Jones getting ready for work each morning can be examined as an example.

Events	**Activities**
1. The alarm rings.	
	A. Jones stirs restlessly.
2. Jones awakens.	
	B. Jones nudges his wife.
	C. Jones lies in bed wishing that he didn't have to go to work.
3. Wife awakens.	
	D. Wife lies in bed wishing that it were Saturday.
4. Jones's wife gets up and begins breakfast.	
Meanwhile	*E.* Wife cooks breakfast.
5. Jones begins morning toilet.	
	F. Jones shaves, bathes, and dresses.
6. The Joneses begin to eat breakfast.	
	G. The Joneses eat part of their breakfast.
7. Jones realizes his bus is about to pass the bus stop.	
	H. Jones jumps up, grabs his briefcase, and runs for the bus.
	I. Wife goes back to bed.
8. Jones boards bus.	
9. Wife falls asleep.	

His PERT network can now be drawn as shown in Figure 12-7. This is a very

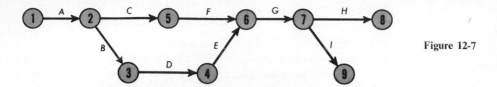

Figure 12-7

elementary example, but it does point up the constituent parts of a PERT network. Note that Jones and his wife must wait until he is dressed (*F*) and the breakfast is cooked (*E*) before they can eat.

In a PERT network each activity should be assigned a specified time for expected accomplishment. The time units chosen should be consistent throughout the network, but the size of the time unit (years, work-weeks, days, hours, etc.) should be selected by the engineer in charge of the project. The time value chosen for each activity should represent the mean of the various times that the activity would take if it were repeated many times.

By using the network of events and activities and by taking into account the times consumed by the various activities, a *critical path* can be established for the project. It is this path that controls the successful completion of the project, and it is important that the engineer be able to isolate it for study. Let us consider the PERT network in Figure 12-8, where the activity times are represented by arabic numbers and are

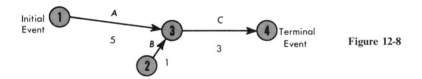

Figure 12-8

indicated in days. Activities represent the expenditure of time and effort. For example, activity *A* (from event 1 to event 3) requires 5 days and is likely devoted to planning the project, while activity *B* requires 1 day and may represent the procurement of basic supplies. Event 1 is the beginning of the project and event 4 is the end of the project. The first step in locating the *critical path* is to determine the "earliest" event times (T_E), the "latest" event times (T_L), and the "slack" time ($T_L - T_E$).

Earliest event times (T_E)

The earliest expected time of an event refers to the time, T_E, when an event can be expected to be completed. T_E for an event is calculated by summing all the activity duration times from the beginning event to the event in question *if the most time-consuming route is chosen*. To avoid confusion, the T_E times of events are usually placed near the network as arabic numbers within rectangular blocks. For reference purposes the beginning of the project is usually considered to be "time zero." In Figure 12-9, T_E for event 3 would be $0 \times 5 = 5$ and T_E for event 4 would be $\boxed{0} + 5 + 3 = \boxed{8}$. However, there are two possible routes to event 4 (*A* + *C*, or *B* + *C*). The *maximum* duration of these event times should be selected as the T_E for event 4. Summing the times, we find

By path *A* + *C:* $\boxed{0} + 5 + 3 = \boxed{8}$ ← Select as T_E for event 4
By path *B* + *C:* $\boxed{0} + 1 + 3 = \boxed{4}$

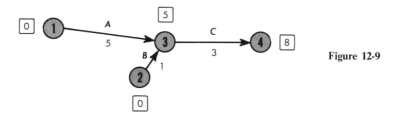

Figure 12-9

Latest event times (T_L)

The latest expected time, T_L, of an event refers to the latest time at which an event
may occur, assuming that the entire project is kept on schedule. T_L for an event is
determined by beginning at the terminal event and working backward through the
various event circuits, subtracting the activity duration *assuming the most time-
consuming route is chosen*. The resulting values of T_L are recorded as arabic numbers
in small ellipses located near the T_E times. Thus, in Figure 12-10, T_L for event
3 would be ⑧ − 3 = ⑤; for event 2, ⑧ − 3 − 1 = ④; and for event
1, ⑧ − 3 − 5 = ⓪.

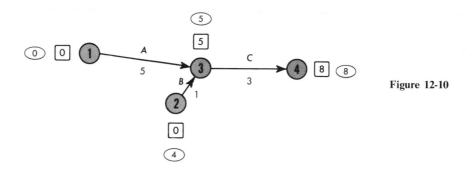

Figure 12-10

Remember that T_L is determined to be the *minimum* of the differences between
the succeeding event T_L and the intervening activity times. Also, in calculating T_L
values one must always proceed backward through the network—from the point of
the arrows to the tail of the arrows.

Slack times

The *slack* time for each event is the difference between the latest event time and the
earliest possible time ($T_L − T_E$). Intuitively, one may verify that it is the "extra time
that an event can slip" and not affect the scheduled completion time of the project.
For example, in Figure 12-14 the slack time for event 2 is ④− $\boxed{0}$ = 4. For this
reason activity B may be started as much as 4 days late and still not cause any overall
delay in the minimum project time of 8 days.

The critical path

The *critical path* through a PERT network is a path that is drawn from the initial event of the network to the terminal event by connecting the events of zero slack. The *critical path* is usually emphasized with a very thick line. Color is sometimes used. In the example problem above the *critical path* would be shown connecting events 1–3–4 (Figure 12-11). Slack times for each event are indicated as small Arabic numbers that are located in triangles adjacent to the events.

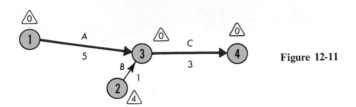

Figure 12-11

Remember that the *critical path* is the path that controls the successful completion of the project. It is also the path that requires the most time to get from the initial event to the terminal event. Any event on the critical path that is delayed will cause the final event to be delayed by the same amount. Conversely, putting an extra effort on noncritical activities will not speed up the project.

Although calculations in this chapter have been done manually, it is conventional practice to program complex networks for solution by digital computer. In this way thousands of activities and events may be considered, and one or more critical paths can be located for further study. Finally, the PERT network should be updated periodically as the work on the project progresses.

The following example will show how a typical PERT diagram is analyzed. It should be noted here, however, that in real-life situations the most difficult task is to identify the precedent relationships that exist and to draw a realistic network of the events and activities. After this is accomplished, following through with a solution technique becomes a relatively routine task.

Example In the PERT network diagram of Figure 12-12, assume that all activity

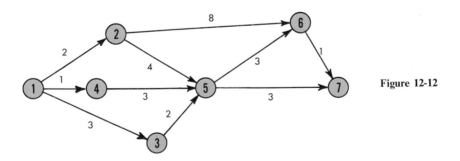

Figure 12-12

times are given in months and that they exist as indicated on the proper activity branch. Find the earliest times, T_E, the latest times, T_L, and the slack times for each event. Identify the critical path through the network.

Solution (See Figure 12-13)

Illustration 12-6

A proper evaluation of PERT will help the engineer to schedule all subcontracts in proper sequence and especially not to allow one work assignment to be pushed ahead of others prematurely or to lag behind unnecessarily.

Figure 12-13

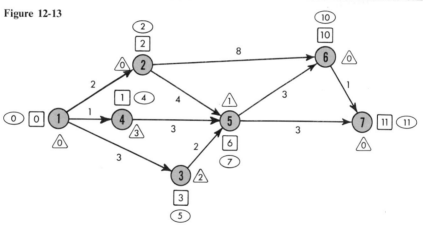

It is usually advisable to construct a summary table of the calculations.

Event ○	Path	T_E □	Path	T_L ○	Slack, $T_L - T_E$ △	On critical path
1	—	0	7–6–2–1	0	0	√
2	1–2	2	7–6–2	2	0	√
3	1–3	3	7–6–5–3	5	2	
4	1–4	1	7–6–5–4	4	3	
5	1–2–5	6	7–6–5	7	1	
6	1–2–6	10	7–6	10	0	√
7	1–2–6–7	11	—	11	0	√

The critical path then is 1–2–6–7 (Figure 12-14). This means that as the project is now organized it will take 11 months to complete.

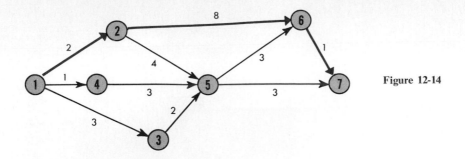

Figure 12-14

Problems

12-1. Consider the network in Figure 12-15. Find T_E, T_L, slack times, and the critical path through the network.

Figure 12-15

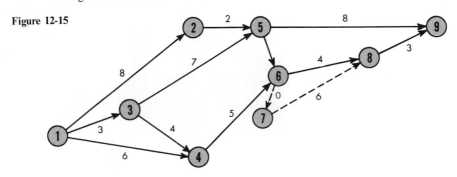

12-2. In Figure 12-15 what effect on project length would the following changes have:
a. Decrease activity 1–2 to 6 days.
b. Decrease activity 5–9 to 1 day.
c. Decrease activity 3–4 to 2 days.

12-3. Explain why "phantom activities" are necessary, and give an example of one.

12-4. Given the following tabular information, determine the PERT network and its critical path.

Activity	Preceding activities	Time
A	None	5
B	None	3
C	A	1
D	B	4
E	B	3
F	E	7

12-5. For some general process with which you are familiar, construct a PERT network. Be sure to label all events and activities.

12-6. Find the critical path in Figure 12-16 and explain its significance here.

12-7. a. Does a decrease in an activity time on the critical path always decrease the project time correspondingly? Why or why not? (*Hint:* See Problem 12-6.)
b. Does an increase in an activity time on the critical path always increase the project time correspondingly? Why or why not? (*Hint:* See Problem 12-4.)

12-8. Given the PERT network in Figure 12-17, when is the earliest possible project completion time?

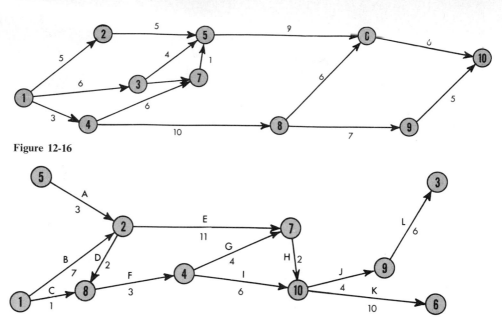

Figure 12-16

Figure 12-17

12-9. If you have extra resources to allocate to one activity in Figure 12-17 where would you put those resources and why? How might this affect the expected duration time of the project?

12-10. One thing that a PERT analysis does not consider is the allocation of limited resources (see Problem 12-9). How can this inability affect the usefulness of a PERT analysis?

Design philosophy

Design philosophy is an important factor with many industries, companies, and consulting firms. The aircraft industry, for example, generally would support a design philosophy that includes (a) lightweight components, (b) safety, (c) limited service life, (d) a wide range of loading conditions and temperature extremes, and (e) concern about vibration and fatigue. The automobile industry, on the other hand, would be more likely to support a design philosophy that stresses (a) consumer price consciousness, (b) long life with minimum service and maintenance, (c) customer appeal, (d) safety for the occupants, and (e) design for mass production.

Some companies are concerned that their products have a "family-like" image and that a responsiveness to customer appeal be designed into all their products. In some instances the image is safety, in some efficiency, in some quality. Public relations should be an important factor with all companies, and the engineer should not be insensitive to the effects that his design will have upon the total company image. The appearance of the product is particularly important in consumer-oriented industries. In such cases the engineer must take this into account in all phases of his design.

Any engineering design is but *one* answer to an identified problem. For this reason few designs have withstood the test of time without undergoing substantial revisions.

Illustration 12-7
Simple designs often have a longer life than complex designs.

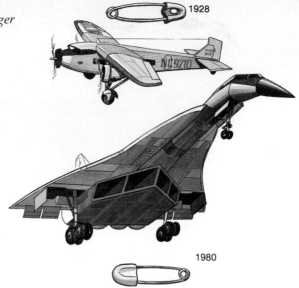

1928

1980

One need but look at the continuous parade of modifications, alterations, changes, and complete redesigns that have taken place within the automobile industry to see how the product of a single industry has been changed thousands and thousands of times. Each change, it was believed at the time, was an improvement over the existing model, even if in appearance only. In some instances this assumption proved to be false, and other modifications were quickly made.

In some situations the pressure for a quick solution has led to the adoption of designs of minimum acceptability. Generally the handicaps and pressures under which the engineer works are of little interest to the customer, who tends to judge the quality of a device or machine by its performance. This emphasis on the product places an additional responsibility upon the shoulders of the engineer to release only those designs which he believes are good designs, those to which he will have no hesitancy in affixing his signature. As a professional person he must be equally aware of his responsibilities to his fellowman and to his employer or client. He must perceive *when he knows,* he must realize *when he does not know,* and he must assume the final responsibility in either case.

Exercises in design

12-11. Estimate the number of drugstores in the United States. Give reasons for your estimate.

12-12. Estimate the number of liters of water of the Mississippi River that pass New Orleans every day. Show your analysis.

12-13. In 100 words or less describe how a household water softener works.

12-14. By the use of simple sketches and a brief accompanying explanation, describe the mechanical operation of a household toilet.

12-15. By the use of a diagrammatic sketch show how plumbing in a home might be installed so that hot water is always instantly available when the hot water tap is opened.

12-16. Analyze and discuss the economic problems involved in replacing ground-level railroad tracks with a suspended monorail system for a congested urban area.

12-17. Discuss the feasibility of railroads offering a service whereby your automobile would be carried on a railroad car on the same train on which you are traveling so that you might have your car available for use upon arrival at your destination.

12-18. Discuss the desirability of assigning an identifying number to each person as soon as they are born. The number could, for example, be tatooed at some place on the body to serve as a social security number, military number, credit card number, and so forth.

12-19. Using local gas utility rates, electric utility rates, coal costs, fuel oil costs, and wood costs, what would be the comparative cost of heating a five-room house in your home community for a winter season?

12-20. Discuss the advantages and disadvantages of having a channel of television show nothing but market quotations, except for brief commercials, during the time the New York stock market and the Chicago commodity market are open.

12-21. You are called to Alaska to consider the problem of public buildings that are sinking in permafrost due to warm weather. What might you do to solve this problem?

12-22. You are located on an ice cap. Ice and snow are everywhere but no water. Fuel and equipment are available. How can you prepare a well from which water can be pumped?

12.23. Assemble the following items: an ink bottle, a marble, a yardstick, an engineer's scale of triangular cross section, five wooden matches, a pocket knife, a candle, a pencil, and a key. Now, using as few of the objects as possible, balance the yardstick across the top "knife-edge" of the engineer's scale in such manner that soon after being released, and without being touched again, it unbalances itself.

12-24. Explain the operation of the rewind mechanism for the hand cord of a home gasoline lawnmower.

12-25. Devise a new method of feeding passengers on airplanes.

12-26. List the consequences of everyone being able to read everyone else's mind.

12-27. At current market values determine the number of years that would be necessary to regain the loss of money (lost salary plus college expense) if one stayed in college one additional year to obtain a master's degree in engineering. What would be the number of years necessary to regain the loss by staying three years beyond the bachelor's degree to obtain a doctorate in engineering?

12-28. Estimate the number of policemen in (a) New York City, and (b) the United States.

12-29. Estimate the number of churches of all faiths in the United States.

12-30. Explain how the following work:
a. An automobile differential.
b. A toggle switch.
c. An automatic cutoff on gasoline pumps.
d. A sewing machine needle when sewing cloth.
e. A refrigeration cycle which does not depend upon electricity.

12-31. With six equal-length sticks construct four equilateral triangles.

12-32. Estimate the number of aspirin tablets now available in the United States.

12-33. A cube whose surface area is 6 mi^2 is filled with water. How long will it take to empty this tank using a 1000 gal/min pump?

12-34. From memory sketch (a) a bicycle, (b) a reel-type lawnmower, (c) a coffee pot, (d) a salt-water fishing reel, and (e) a rifle.

12-35. Make something useful from the following items: a piece of corrugated cardboard 12 in. $\times$ 24 in., 6 ft of string, 3 pieces of chalk, 10 rubber bands, a small piece of gummed tape, 3 tongue depressors, 5 paper clips, and 7 toothpicks.

12-36. Propose some way to eliminate the need for bifocal glasses.

12-37. Design a device that can measure to a high degree of accuracy the wall thickness of a long tube whose ends are not accessible.

12-38. Design a man's compact travel kit that can be carried in the inside coat pocket.

12-39. Design a home-type sugar dispenser for a locality where the average rainfall is 100 in./yr.

12-40. Design a new type of men's apparel to be worn around the neck in lieu of a necktie.

12-41. Design a new type of clothespin.

12-42. Design a new fastener for shirts or blouses.

12-43. Design a personal monogram.

12-44. Design a device to aid federal or civil officers in the prevention or suppression of crime.

12-45. Design a highway system and appropriate vehicles for a country where gasoline is not obtainable and where motive power must be supplied external to the vehicle.

12-46. Design an electrical system for a home that does not receive its energy from a power company or a storage battery.

12-47. Design a device for weighing quantities of food for astronauts who are enroute to the moon.

12-48. Design a machine or process to remove Irish potato peelings.

12-49. Design a "black-eyed pea" sheller.

12-50. Design a corn shucker.

12-51. Design a trap to snare mosquitoes alive.

12-52. Design the "ideal" bathroom, including new toilet fixtures.

12-53. Design a toothpaste dispenser.

12-54. Design a woozle.

12-55. Design a device that would enable paralyzed people to read in bed.

12-56. Design a jiglike device that an amateur "do-it-yourself" home workman could use to lay up an acceptably straight brick wall.

12-57. Design a device to retail for less than $20.00 to warn "tailgaters" that they are too close to your automobile.

12-58. Devise a system of warning lights connected to your automobile that will warn drivers in cars following you of the changes in the speed of your car.

12-59. You live in a remote community near the Canadian border, and you have a shallow well near your home from which you can get a copious supply of water. Although the water is unfit for drinking or irrigation, its temperature is a constant 64°F. Design a system to use this water to help heat your home.

12-60. Design and build a prototype model of a small spot welder suitable for use by hobby craftsmen. Prepare working sketches and make an economic study of the advisability of producing these units in volume production.

12-61. Design some device that will awaken a deaf person.

12-62. Design a coin-operated hair-cutting machine.

12-63. Design a two-passenger battery-powered Urbanmobile for use around the neighborhood, for local shopping center visits, to commute to the railway station, and so on. The rechargeable battery should last for 60 mi on each charge. Provide a complete report on the design, including a market survey and economic study.

12-64. Design some means of visually determining the rate of gasoline consumption (mi/gal) at any time while the vehicle is in operation.

12-65. Design a device to continuously monitor and/or regulate automobile tire pressures.

12-66. Design a novel method of catching and executing mice that will not infringe the patent of any other known system now on the market.

12-67. Design a new toy for children ages 6 to 10.

12-68. Design a device to replace the conventional oarlocks used on all rowboats.

12-69. Devise an improved method of garbage disposal for a "new" city that is to be constructed in its entirety next year.

12-70. Design and build a simple device to measure the specific heat of liquids. Use components costing less than $3.00.

12-71. Design for teenagers an educational hobby kit that might foster an interest in engineering.

12-72. Design a portable traffic signal that can be quickly put into operation for emergency use.

12-73. Design an egg breaker for kitchen use.

12-74. Design an automatic dog-food dispenser.

12-75. Design a device to automatically mix body soap in shower water as needed.

12-76. Design an improved keyholder.

12-77. Design a self-measuring and self-mixing epoxy glue container.

12-78. Design an improved means of cleaning automobile windshields.

12-79. Design a noise suppressor for a motorcycle.

12-80. Design a collapsible bicycle.

12-81. Design a tire-chain changer.

12-82. Design a set of improved highway markers.

12-83. Design an automatic oil-level indicator for automobiles.

12-84. Design an underwater means of communication for skin divers.

12-85. Design a means of locating lost golf balls.

12-86. Design a musician's page turner.

12-87. Design an improved violin tuning device.

12-88. Design an attachment to allow a motorcycle to be used on water.

12-89. Design a bedroll heater for use in camping.

12-90. Design an improvement in backpacking equipment.

12-91. Design an improved writing instrument.

12-92. Design a means of disposing of solid household waste.

12-93. Design a type of building block that can be erected without mortar.

12-94. Design a means for self-cleaning of sinks and toilet bowls.

12-95. Design some means to replace door knobs or door latches.

12-96. Design a simple animal-powered irrigation pump for use in developing nations.

12-97. Design a therapeutic exerciser for use in strengthening weak or undeveloped muscles.

12-98. Design a Morse-code translator that will allow a deaf person to read code received from radio receivers.

12-99. Design an empty-seat locator for use in theaters.

12-100. Design a writing device for use by armless people.

12-101. Design and build an indicator to tell when a steak is cooked as desired.

12-102. Design a device that would effectively eliminate wall outlets and cords for electrical household appliances.

12-103. Design the mechanism by which the rotary motion of a 1-in.-diameter shaft can be transferred around a 90° corner and imparted to a $\frac{1}{2}$-in.-diameter shaft.

12-104. Design a mechanism by which the vibratory translation of a steel rod can be transferred around a 90° corner and imparted to another steel rod.

12-105. Design a device or system to prevent snow accumulation on the roof of a mountain cabin. Electricity is available, and the owner is absent during the winter.

12-106. Using the parts out of an old spring-wound clock, design and fabricate some useful device.

12-107. Out of popsicle sticks build a pinned-joint structure that will support a load of 50 lb.

12-108. Design a new device to replace the standard wall light switch.

12-109. Design and build a record changer that will flip records as well as change them.

12-110. Design a wheelchair that can lift itself from street level to a level 1 ft higher.

12-111. Design a can opener that can be used to make a continuous cut in the top of a tin can whose top is of irregular shape.

12-112. Design and build for camping purposes a solar still that can produce 1 gallon of pure water per day.

12-113. Design, build, and demonstrate a device that will measure and indicate 15 seconds of time as accurately as possible. The device must not use commercially available timing mechanisms.

12-114. Few new musical instruments have been invented within the last 100 years. With the availability of modern materials and processes, many novel and innovative designs are now within the realm of possibility. To be marketable over an extended period of time such an instrument should utilize the conventional diatonic scale of eight tones to the octave. It could, therefore, be utilized by symphonies, in ensembles, or as a solo instrument using existing musical compositions. You are the chief engineer for a company whose present objective is to create and market such a new instrument. Design and build a prototype of a new instrument that would be salable. Prepare working drawings of your model together with cost estimates for volume production of the instrument.

12-115. Design some means of communicating with a deaf person who is elsewhere (such as by radio).

12-116. For a bicycle, design an automatic transmission that will change gears according to the force applied.

12-117. Design a "decommercializer" that will automatically cut out all TV commercial sounds for 60 sec.

12-118. Design a solar-powered refrigerator.

12-119. Design a small portable means for converting seawater to drinking water.

12-120. Design a fishing lure capable of staying at any preset depth.

12-121. Design an educational toy that may be used to aid small children in learning to read.

12-122. Design some device to help a handicapped person.

12-123. Design a heating and cooling blanket.

12-124. Design a portable chair.

12-125. Design a portable solar cooker.

12-126. Design a carbon monoxide detector for automobiles.

12-127. Design a more effective method for prevention and/or removal of snow and ice from military aircraft.

12-128. Design a "practical" vehicle whose operation is based upon the "ground-effect" phenomenon.

12-129. Design a neuter (neither male nor female) connector for quick connect and disconnect that can be used on the end of flexible hose to transport liquids.

12-130. Design an electric space heater rated from 10,000 Btu/hr to 50,000 Btu/hr for military use in temporary huts and enclosures.

12-131. There is need for a system whereby one device emplaced in a hazardous area (minefield or other denial area) would interact with another device issued to each soldier, warn him of danger, and send guidance instructions for him to avoid or pass through the area of safety. Design such a system.

12-132. Develop some method to rate and/or identify the presence of rust spots when coatings fail to protect metal adequately. Present visual methods are unreliable and variable in results.

12-133. Develop a system whereby diseases of significance could be diagnosed rapidly and accurately.

12-134. Design a strong, flexible, lumpless, V-belt connector.

12-135. Design an inexpensive system for keeping birds out of ripening fruit trees.

12-136. Design a replacement for the paper stapler which will not puncture the paper.

12-137. Design and construct a vehicle that will carry a payload across the classroom floor as far as possible.

Specifications
  1. The vehicle is to be powered by a conventional spring-activated household mousetrap about 4.75 cm by 10 cm.
  2. The maximum dimensions of the vehicle are 20 cm long, 15 cm wide, and 15 cm high.
  3. Payload must be in one piece and removable for weighing. The maximum dimensions for the payload are 5 cm $\times$ 5 cm $\times$ 5 cm.
  4. The amount of the payload to be carried, materials of construction, and vehicle design are your responsibility.

Testing
  A maximum of 1 minute will be allowed to position the vehicle and prepare it for

the run. The vehicle must cover at least a distance of 5 meters or it will be disqualified.

Evaluation

A total of five quantities are included in the evaluation. They are:

P = payload weight (newtons)
L = total distance traversed by the vehicle from the starting line (meters)
T = total elapsed time to traverse 5 meters (seconds)
W = vehicle weight (newtons)
C = cost of the vehicle at $0.10/newton

The overall value of the vehicle will be determined by the following formula:

$$V = \frac{P \times L^2}{T \times W \times C}$$

To complete the individual design project, after testing prepare a report containing the calculation of the value of your vehicle, a copy of the preliminary sketch, and a brief analysis of your reasoning for the particular design you used in the construction. Discuss the advantages and limitations of your design. How might the design be improved?

12-138. Design and construct a powered, self-controlled surface vehicle that will negotiate a "figure 8" course on a smooth, horizontal surface whose dimensions are 1 m × 2 m.

Specifications

Construct the vehicle from the following materials:
1. Balsa wood and/or cardboard ≤5.0 mm thick, not to exceed 1500 cm².
2. Cotton thread (no nylon!) ≤20 gage, not to exceed 30 cm in length.
3. Balsa wood cement; not more than one small tube.
4. Maximum of four standard-size paper clips.
5. Maximum of four circular rubber bands. The original width of each band must not exceed 4 mm, and the unstretched length of the elongated oval must not exceed 10 cm.

Cost Schedule

		Quantity	Cost
Balsa wood or cardboard	at $1.00/cm³		
Cotton thread	at $3.00/cm		
Rubber bands	at $50.00 each		
Standard office paper clips	at $20.00 each		
	Total cost		

Evaluation

Three quantities are to be evaluated in determining the value (V) of the vehicle. They are:

W = weight of vehicle (newtons)
C = cost of vehicle (dollars)
f = fraction of "figure 8" course successfully negotiated

$$V = \frac{f \times 10^6}{W + C}$$

To complete the individual design project, after testing prepare a report containing the calculation of the value of your vehicle, a copy of the preliminary sketch, and a brief analysis of your reasoning for the particular design you used in the construction. Discuss the advantages and limitations of your design. How might the design be improved?

12-139. Design a bridge to span 450 mm between supports, with a 30-mm-wide roadway and having a vertical road clearance of at least 20 mm. (Suspension type bridges must have an actual roadway and vertical "tower" supports.) Construct the bridge to support a load that will be applied to the center of its roadway, midway between the two supports.

Construction Materials
Bridge construction materials are limited to the following:

Ordinary soda straws	at $0.01/mm
Plain cardboard (maximum thickness 2 mm)	at $0.01/m²
Corrugated cardboard (maximum thickness 6 mm)	at $0.02/m²
Cotton string (no nylon)	at $0.01/mm
Ordinary round, wooden toothpicks	at $0.50 each
Standard office paper clips	at $1.00 each

Quantity	Cost
Total cost	

Note: Adhesive paste, glue, epoxy, casein, etc. (no tape) can be used to join the materials at no charge. However, this material will serve only the function of fasteners to connect structural members.

Evaluation
Three quantities are to be evaluated in determining the value (V) of the bridge. They are:

W = weight of the bridge (newtons)
L = load applied at center of bridge that will deflect the center of the bridge 40 mm—or to failure, if that occurs first (newtons)
C = cost of construction materials (dollars)

$$V = \frac{L \times 10^2}{W + C}$$

To complete the individual design project, after testing prepare a report containing the calculation of the value of your bridge, a copy of the preliminary sketch, and a brief analysis of your reasoning for the particular design you used in the construction. Discuss the advantages and limitations of your design. How might the design be improved?

12-140. Design a nozzle or device that will utilize a round rubber balloon to drive a boat, transporting a mass of your choice, selected from those made available by the instructor. The channel of water used to float the boat is 5 in wide, 2 in. deep, and 120 in. long.

Boat Construction

The boat may be made of any material and is to be constructed as shown in Figure 12-18. The balloon shall be a new, ordinary rubber balloon. The boat may not be altered in any way (including driving nails or wires into the surface). However, tape, rubber bands, etc., may be used to attach a balloon supporting structure. The propulsion device must be an original design. It will be attached to the boat via the 1 in. $\times$ 2 in. $\times \frac{1}{2}$ in. slot on the stern.

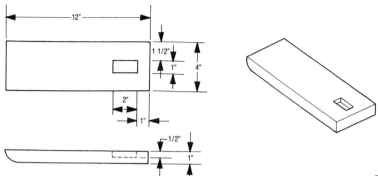

Figure 12-18

Testing

The balloon must be inflated to 7 in. diameter at the start of the test (use a go–no-go wire gage for measurement). Allow a maximum of 2 minutes for a test run, including preparation for test and actual test. The boat may be placed anywhere in the test trough at start. Use a pencil as a gate to block the boat from moving prematurely. Air is to be leaving balloon before the gate is lifted.

Evaluation

Three quantities are to be evaluated in determining the design performance (P). They are:

M = (number of steel washers carried) $\times$ 20/980 g$_m$
L = distance (cm) front of boat travels from time "zero" to time T
T = time (sec) from point where gate is lifted to point when all of the air is exhausted from the balloon, or when the boat strikes the end of the trough—whichever occurs first.

$$P = \frac{ML}{T}$$

To complete the individual design project, after testing prepare a report containing the calculation of the value of your boat, a copy of the preliminary sketch, and a brief analysis of your reasoning for the particular design you used in the construction. Discuss the advantages and limitations of your design. How might the design be improved?

12-141. Design a structure that will fit around an object having a parabolic profile. The specific dimensional constraints are as follows:

    A The structure must be designed so that a parabolic profile, 4 in. $\times$ 6 in., shown at A in Figure 12-19, may pass completely through the space between the structure and the loading cable plane.[5]

B, C, D Overall dimensions of the structure cannot exceed 10 in. $\times$ 10 in. $\times$ 4 in.

[5] Adapted from a problem used at Carnegie Mellon University.

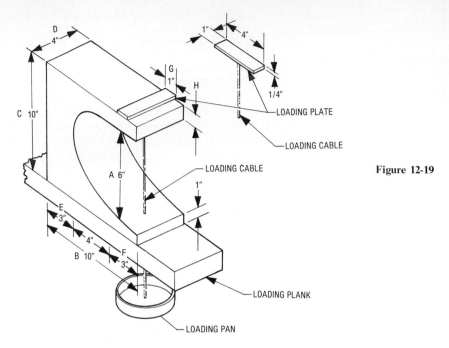

Figure 12-19

    E Maximum thickness of the structure cannot exceed 3 in.
    F Extension of the base of the structure beyond the loading cable cannot exceed 3 in.
    G Overhand of the top of the structure beyond the loading cable cannot exceed 1 in.
  H, I Dimensions H and I are left to the discretion of the designer, provided that they do not make the overall height of the structure exceed 10 in.

Materials

Only balsa wood and glue may be used as materials. Maximum cross section for sticks is $\frac{3}{4}$ in. $\times$ 1 in. Maximum cross section for sheets is $\frac{3}{8}$ in. $\times$ 4 in. The use of glue will be restricted to areas actually joining two pieces of wood. The wood, other than at joints, may not be coated or impregnated with lacquer, shellac, paper, or any other material. The use of glued laminations of sheets of balsa wood is prohibited. There may be no concealed elements in the structure.

Testing

The load will be applied in a vertical direction by a cable attached to the center of a 4 in. $\times$ 1 in. $\times$ $\frac{1}{4}$ in. steel loading plate, which will rest on top of the structure. The loading cable will pass through an opening provided in the base of the structure and through an opening in the loading plank. The base of the structure must be self-supporting and rest on the loading plank. The top of the structure must be designed to accommodate the loading plate with a cable attached so that it may be easily inserted or placed on the structure without the use of any clamps, screws, glue, etc. Failure under load will be considered to be the point where the structure collapses, can no longer carry additional load, or the specified parabolic profile will no longer clear the structure and the plane of the loading cable. The load carried by the structure will include the weight of the loading plate and cable.

Evaluation

The quality of performance (P) of the structure will be determined by two factors, as follows:

L — load necessary for failure (lb)

W = weight of structure (lb)

$$P = \frac{L}{W}$$

To complete the individual design project, after testing prepare a report containing the calculation of the value of your structure, a copy of the preliminary sketch, and a brief analysis of your reasoning for the particular design you used in the construction. Discuss the advantages and limitations of your design. How might the design be improved?

12-142. Design, construct, and test a wind-powered unit that will lift a stationary $\frac{1}{3}$-kg weight a distance of 1 meter (Figure 12-20). The wind source is a medium-sized electric fan, not to exceed $\frac{1}{8}$ hp. Your design should have a base capable of being held on the table with a laboratory C-clamp. The fan will be positioned 1 meter from the edge of the test table, as shown. Your design must receive its energy from the air movement of the fan.

Construction Materials

Any materials may be selected for your design, including the type of string that will be used as the lift cable. However, you cannot use any industrial-fabricated components for your design.

Figure 12-20

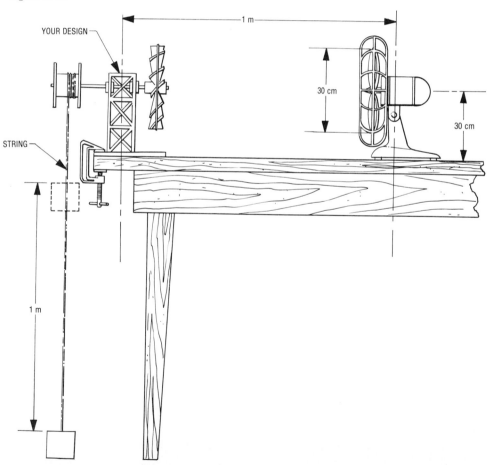

Evaluation

The design performance (P) depends upon three factors, as follows:

h = height of weight lifted (meters

t = time of the lift (seconds)

w = weight of the design, including string (newtons)

$$P = \frac{h}{tw}$$

To complete the individual design project, after testing prepare a report containing the calculation of the value of your unit, a copy of the preliminary sketch, and a brief analysis of your reasoning for the particular design you used in the construction. Discuss the advantages and limitations of your design. How might the design be improved?

Appendixes

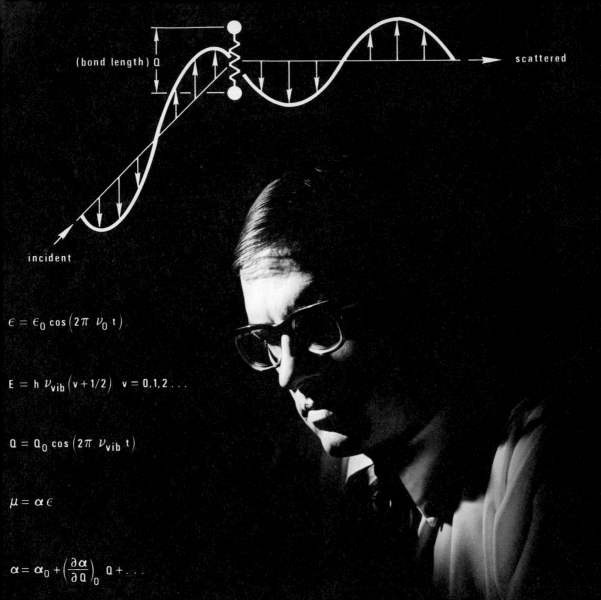

$$\epsilon = \epsilon_0 \cos \left(2\pi \, \nu_0 \, t \right)$$

$$E = h \, \nu_{\mathbf{vib}} \left(v + 1/2 \right) \quad v = 0,1,2 \ldots$$

$$Q = Q_0 \cos \left(2\pi \, \nu_{\mathbf{vib}} \, t \right)$$

$$\mu = \alpha \, \epsilon$$

$$\alpha = \alpha_0 + \left(\frac{\partial \alpha}{\partial Q} \right)_0 Q + \ldots$$

(RAYLEIGH)   (STOKES RAMAN)   (ANTISTOKES RAMAN)

$$\mu = \epsilon_0 \, \alpha_0 \cos \left(2\pi \, \nu_0 \, t \right) + \frac{\epsilon_0}{2} \left(\frac{\partial \alpha}{\partial Q} \right)_0 Q_0 \cos \left[2\pi \left(\nu_0 - \nu_{\mathbf{vib}} \right) t \right] + \frac{\epsilon_0}{2} \left(\frac{\partial \alpha}{\partial Q} \right)_0 Q_0 \cos \left[2\pi \left(\nu_0 + \nu_{\mathbf{vib}} \right) t \right]$$

APPENDIX I

A summary of technical drawing fundamentals

Technical drawings may be drawn formally with instruments, or they may be sketched freehand using only a pencil or some other marking tool. Freehand sketches are often used in developing design concepts and in planning formal drawings. Instrument drawings are normally required when it is necessary to provide accurate descriptions for manufacturing or construction.

AI.1 Drawing instruments

When making instrument drawings, specialized tools are used to draw the different types of lines required to describe an object.

T-squares, parallel rules, and various types of drafting machines ... used to establish a horizontal frame of reference and to draw horizontal lines.

Figure AI-1

Parallel Rule

Track Type Drafting Machine

Figure AI-2

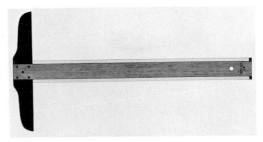

T-Square

Triangles . . . used with T-squares and parallel rules to draw vertical and angled lines. These lines can be drawn directly with drafting machines.

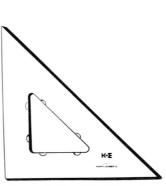

Figure AI-3

45° and 30°–60° Triangles

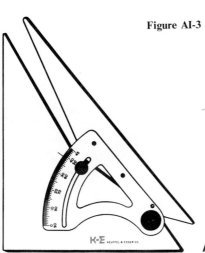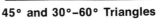

Adjustable Triangle

Compasses, templates, and irregular curves
... used to draw curved lines.

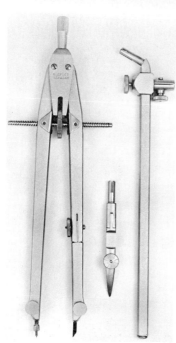

Figure AI-4

Compass and Accessories

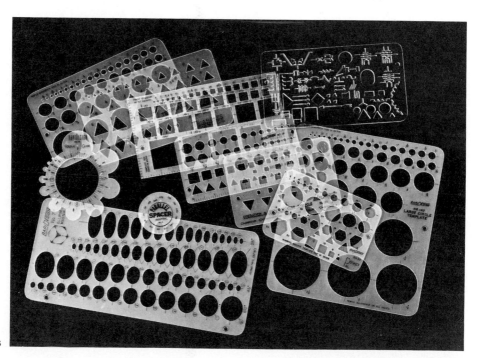

Templates

Irregular Curve

Dividers and scales ... used to measure or lay out line lengths.

Dividers

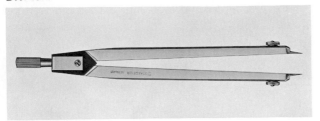

Figure AI-5

Scale

Protractors ... used to measure or lay out angles.

Figure AI-6

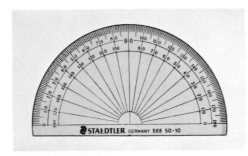

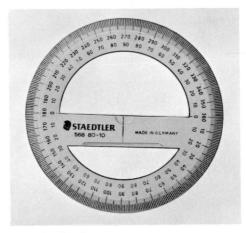

180° Protractor   **360° Protractor**

Pencils ... used to draw lines. Many types available varying according to lead holding method, sharpening, lead diameter, and type and hardness of lead. Rotate pencil slowly while drawing lines to maintain conical point.

Ultra-Thin Mechanical Pencil

Lead Holder

Wood Pencil

Figure AI-7

Lead Grades

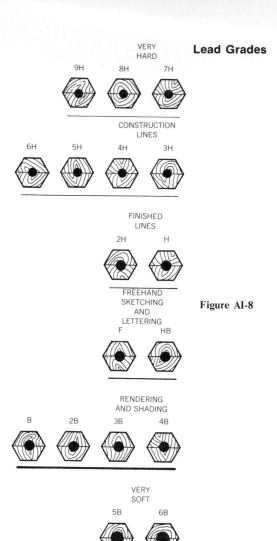

VERY
HARD

9H   8H   7H

CONSTRUCTION
LINES

6H   5H   4H   3H

FINISHED
LINES

2H   H

FREEHAND
SKETCHING
AND
LETTERING

F   HB

RENDERING
AND SHADING

B   2B   3B   4B

VERY
SOFT

5B   6B

Figure AI-8

Technical pens ... used alone or in compasses to draw lines. Keep capped between
uses to prevent clogging. Clean frequently.

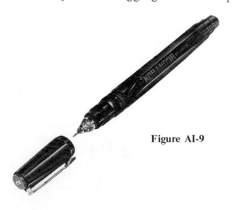

Figure AI-9

INK LINES

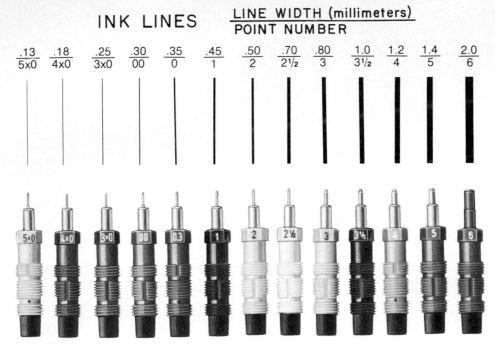

LINE WIDTH (millimeters)
POINT NUMBER

| $\frac{.13}{5\times0}$ | $\frac{.18}{4\times0}$ | $\frac{.25}{3\times0}$ | $\frac{.30}{00}$ | $\frac{.35}{0}$ | $\frac{.45}{1}$ | $\frac{.50}{2}$ | $\frac{.70}{2\frac{1}{2}}$ | $\frac{.80}{3}$ | $\frac{1.0}{3\frac{1}{2}}$ | $\frac{1.2}{4}$ | $\frac{1.4}{5}$ | $\frac{2.0}{6}$ |

Figure AI-10

AI.2   Line drawing procedures

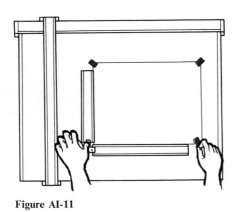

Figure AI-11

The drawing paper is lined up horizontally . . . corners are taped to the board.

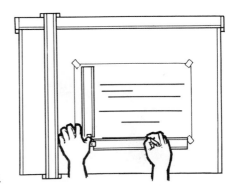

Figure AI-12

Horizontal lines are drawn directly.

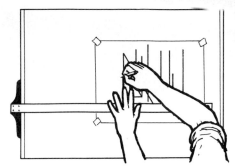

Vertical lines are drawn with a triangle placed on a T-square or parallel rule....

Figure AI-13

or drawn directly with a drafting machine.

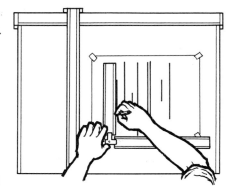

Figure AI-14

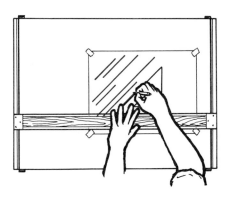

Angled lines are also drawn with a triangle

Figure AI-15

...or directly with a drafting machine.

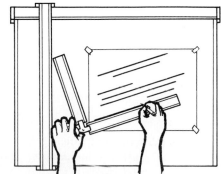

Figure AI-16

Circles must be located first with centerlines
. . . then drawn with a compass or template.

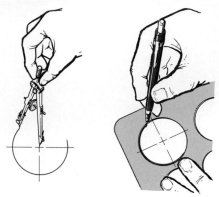

Figure AI-17

Irregular curves are located by plotting
points . . . the curve is lightly sketched . . .
then darkened in with an irregular curve.

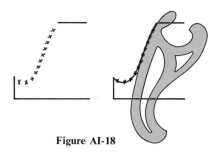

Figure AI-18

Parallel lines are drawn by sliding triangles
on one another or on a T-square or parallel
rule.

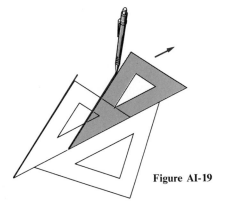

Figure AI-19

Figure AI-20

Perpendicular lines can be drawn by sliding
triangles on one another or on a T-square or
parallel rule.

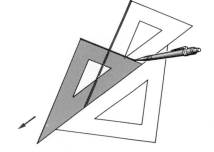

Perpendicular lines can be constructed with a compass.

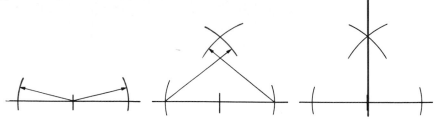

Figure AI-21

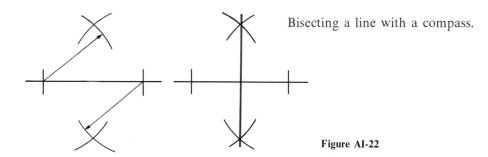

Bisecting a line with a compass.

Figure AI-22

Dividing a line into parts using existing scales ... set the scale at a convenient angle and draw parallel lines.

Figure AI-23

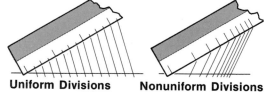

Uniform Divisions   **Nonuniform Divisions**

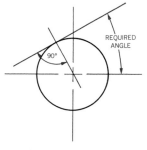

Drawing a line tangent to a circle or an arc ... the line is perpendicular to a radial line drawn to the point of tangency.

Figure AI-24

Drawing regular geometric shapes ... draw a circle first and then draw tangent lines with a triangle or drafting machine.

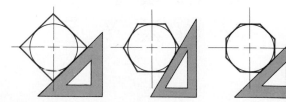

Figure AI-25

Drawing ellipses ... lay out a rhombus with sides equal to the diameter of the circle from which the ellipse is derived ...

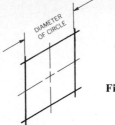

Figure AI-26

Figure AI-27

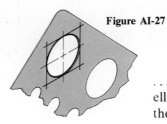

... use the rhombus as a guide in selecting an ellipse template opening that is tangent to the four sides at their midpoints ...

... or draw perpendiculars from the mid-points of the sides to locate centers for compass arcs that will approximate the ellipse.

Figure AI-28

AI.3 Lettering

Standard letter and number forms are used for clarity, repeatability, and ease of drawing. The customary sequence of pencil strokes is labeled 1, 2, 3, ...

Capital Letters:

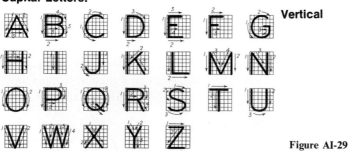

Vertical

Figure AI-29

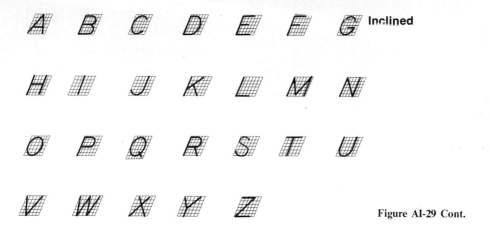

Figure AI-29 Cont.

Numbers:

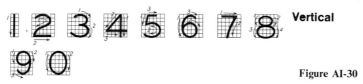

Vertical

Figure AI-30

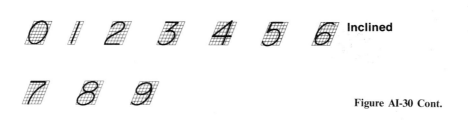

Inclined

Figure AI-30 Cont.

Lower Case Letters:

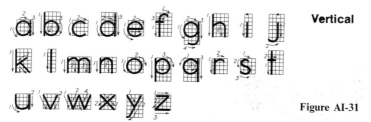

Vertical

Figure AI-31

Inclined

Figure AI-31 Cont.

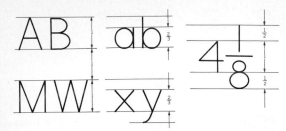

Figure AI-32

Lettering technique ... draw light guidelines. Most engineering lettering is 3 mm (0.12 in.) high with proportions as shown.

Maintain uniform slant. Do not retrace letters. Rotate pencil slightly between letters to maintain its conical point.

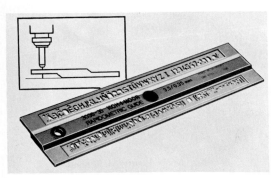

Figure AI-33

Letter spacing ... maintain equal area between letters.

ONE LETTER. WORDS

Figure AI-34

Word spacing ... allow one letter width between words and after punctuation.

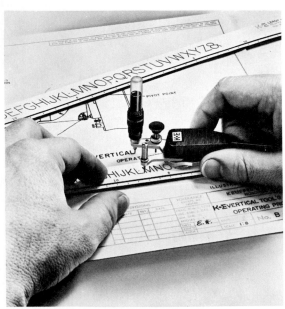

Ink lettering ... used where greater clarity and reproducibility is required. Especially required if drawing will be reduced in size when it is duplicated. For best quality, the pen is guided by a template or held in a scriber that follows a template.

Figure AI-35

Transfer lettering ... used in special situations where large letters and special letter forms are needed. Individual letters are pressed or rubbed onto drawing surface.

Figure AI-36

AI.4   Multiview drawing

Multiview drawings are made using the orthographic projection system. In this system, the drawing on paper is the picture that the viewer sees on the surface of a transparent viewing plane placed perpendicular to the line of sight between the viewer's eyes and the object.

Figure AI-37

ORTHOGRAPHIC—LINES OF SIGHT ARE PARALLEL AND PERPENDICULAR TO THE VIEWING PLANE.

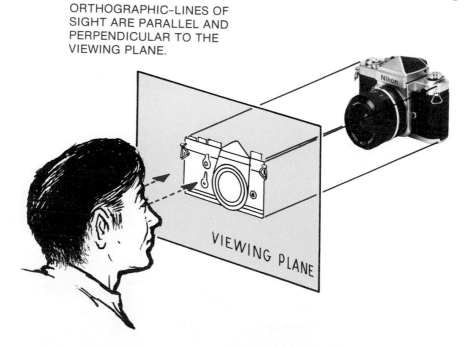

Principal views ... by placing the object in different positions relative to the viewing plane, a variety of views may be recorded. Principal views are obtained by positioning the object's principal faces—front and rear, top and bottom, left and right sides—parallel to the viewing plane.

Views are aligned with each other and arranged in the standard order shown so that they can be quickly recognized. Usually two or three principal views are sufficient unless the object is quite complex.

TOP VIEW

REAR VIEW   LEFT SIDE VIEW   FRONT VIEW   RT. SIDE VIEW

BOTTOM VIEW

Figure AI-38

Layout of views ... the size and shape of an object are described using a rectangular coordinate measuring system. The dimensions of space are width, height, and depth.

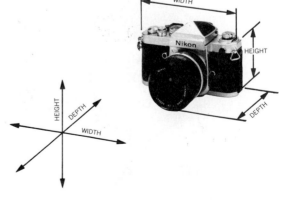

Figure AI-39

Each orthographic view shows only two dimensions of space.

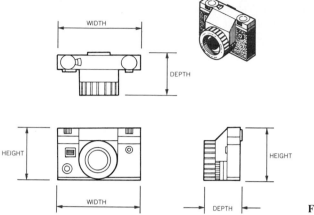

Figure AI-40

Lines and surfaces cannot disappear . . . they must be accounted for in all views. Lines may appear true length, forshortened, or as a point.

Planes may appear true size, distorted, or as a line. Even if distorted, the characteristic shape is retained . . . rectangles have four sides that are parallel in all views.

Figure AI-41

Selection of views ... the object is positioned so that its most characteristic contour is seen in the "front" view position. Draw only the minimum number of views required to completely describe the object.

The views selected should be those that show the shape of the features most clearly with the fewest hidden lines.

Figure AI-42

Types of lines ... a standard "alphabet" of lines is used to describe an object on a drawing. Object lines outline the visible features. Any feature lines that are hidden are drawn using the hidden-line symbol.

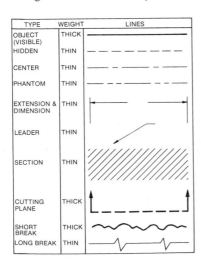

TYPE	WEIGHT	LINES
OBJECT (VISIBLE)	THICK	
HIDDEN	THIN	
CENTER	THIN	
PHANTOM	THIN	
EXTENSION & DIMENSION	THIN	
LEADER	THIN	
SECTION	THIN	
CUTTING PLANE	THICK	
SHORT BREAK	THICK	
LONG BREAK	THIN	

Figure AI-43

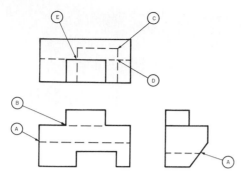

Figure AI-44

Hidden lines start and end with a dash Ⓐ ... except when the hidden line is a continuation of a visible line Ⓑ.

Intersecting hidden lines meet at a dash Ⓒ ... nonintersecting hidden lines do not touch Ⓓ.

Hidden and visible lines that cross do not touch Ⓔ.

Centerlines are drawn on all views of cylindrical objects. The centerlines are broken between views.

Figure AI-45

AI.5   Auxiliary views

In multiview drawing, auxiliary views are produced when the principal faces of the object are positioned at an angle to the viewing plane. The viewer's line of sight is no longer perpendicular to a principal face. Only the line or surface to which the viewing plane is parallel is seen in its true size. By directing lines of sight at appropriate angles, auxiliary viewing planes perpendicular to the line of sight can be established. Any line, surface, angle, or distance to which the auxiliary viewing plane is parallel will appear true size in the auxiliary view that is produced.

Viewing directions

Principal view ... the line of sight is perpendicular to one of the principal faces of the object.

Figure AI-46

Primary auxiliary view ... the line of sight is at an angle to two principal faces of the object (top and front in this example).

Figure AI-47

Secondary auxiliary view . . . the line of sight is at an angle to three principal faces of the object (top, front, and right side in this example).

Figure AI-48

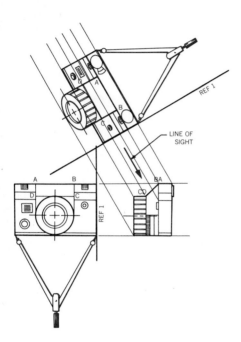

Figure AI-49

Layout of auxiliary views

For a primary auxiliary . . . direct the line of sight at an angle to a principal view . . . draw projection lines parallel to the line of sight from the features of the principal view.

Establish reference line 1 perpendicular to the projection lines at a convenient location . . . in a view adjacent to the one from which you projected (the front view in this example), draw a mating reference line 1 perpendicular to its projection lines.

Plot points defining the features of the object by transferring distances from reference lines 1 . . . label the points to prevent error and to save time.

The example shows a line of sight directed perpendicular to an inclined plane that is seen on edge in the side view. This establishes an auxiliary viewing plane parallel to the inclined plane. The auxiliary view shows the true size of this plane and the true lengths of its lines.

For a secondary auxiliary . . . direct a line of sight at an angle to a principal view and draw a primary auxiliary view.

Direct a second line of sight at any angle to this view . . . draw projection lines parallel to the line of sight from the features of the primary view.

Establish reference line 2 perpendicular to the secondary projection lines at a convenient location . . . in the view adjacent to the one from which you projected (the front view in this example), draw a mating reference line 2 perpendicular to the projection lines used to draw the primary auxiliary view.

Plot points defining the features of the object by transferring distances from reference lines 2. In this example the secondary auxiliary view shows a pictorial type view of the entire object.

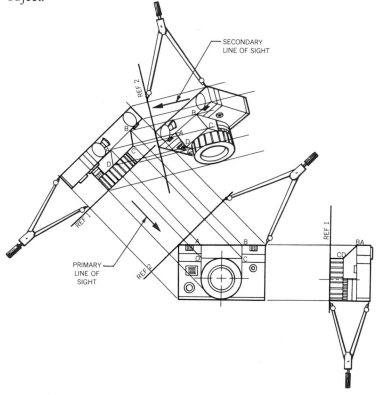

Figure AI-50

AI.6 Section views

Section views show features that are hidden inside an object. The layout procedure for a section view is the same as that for any other type of view in multiview drawing. The object is imagined to be cut open at a selected location with the viewing plane positioned parallel to the cut surface.

Layout of section views

Cutting plane line . . . shows location of cut and direction of viewing. The section view is placed behind the arrows.

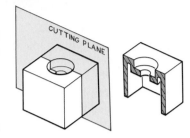

Figure AI-51

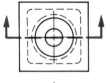

Section lining . . . shows the cut surface. Hidden lines are normally omitted in section views.

Figure AI-52

When several section views are drawn, or when a view must be moved out of its normal projected location, each section must be labeled.

A view must retain its proper orientation when moved and must remain behind the arrows. The section lining pattern remains constant throughout the drawing.

Figure AI-53

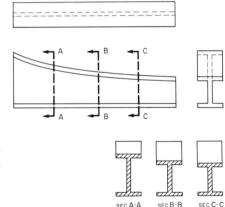

SEC A-A   SEC B-B   SEC C-C

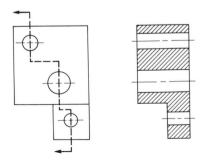

Special types of section views

Offset section . . . the cutting plane passes through several features to reduce the number of views required. The changes in direction of the cutting plane are not shown on the section view.

Figure AI-54

Aligned section ... symmetrical features may be revolved to avoid a distorted appearance in the section view.

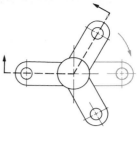

Figure AI-55

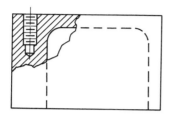

Broken-out section ... a view may be partially sectioned to show only a small feature. No cutting plane is shown.

Figure AI-56

Revolved section ... the cross-sectional shape of certain features can be shown directly on a view. The cut is assumed to be at the centerline of the section view, so no cutting plane is needed.

When a revolved section is moved outside of the view, it is called a removed section.

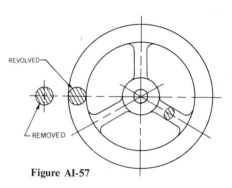

REVOLVED

REMOVED

Figure AI-57

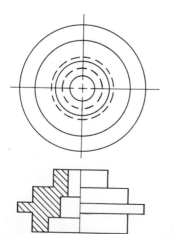

Half section ... on a symmetrical object both inside and outside features can be shown in the same view. The cutting plane is not shown because it is always assumed to pass through the axis of symmetry. Hidden lines are often omitted in the "outside" half since the section portion shows the inside contours.

Figure AI-58

Assembly section ... each cut part of an assembled object must be given a distinguishing section lining pattern that remains constant throughout the drawing. The patterns on adjacent parts must contrast in angle and/or spacing.

Shafts and standard parts are not sectioned when on the cutting plane.

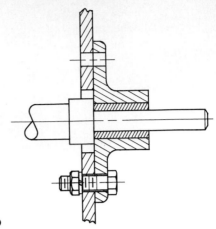

Figure AI-59

AI.7 Threads and fasteners

Fasteners and the holes associated with them frequently appear on engineering drawings. To save time, items such as these are usually drawn in a simplified or symbolic manner.

Standard threads ... drawn symbolically unless of very large diameter. The nominal diameter of the thread is drawn to scale. The lines representing the threads are spaced for realistic appearance only.

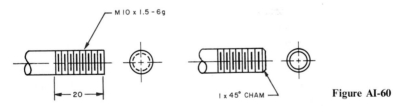

M 10 x 1.5 - 6g

20

1 x 45° CHAM

Figure AI-60

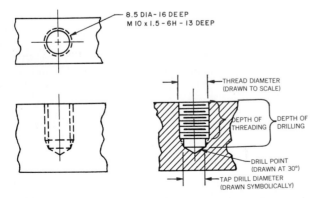

8.5 DIA - 16 DEEP
M 10 x 1.5 - 6H - 13 DEEP

THREAD DIAMETER (DRAWN TO SCALE)

DEPTH OF THREADING

DEPTH OF DRILLING

DRILL POINT (DRAWN AT 30°)

TAP DRILL DIAMETER (DRAWN SYMBOLICALLY)

Figure AI-61

Common fasteners ... drawn slightly simplified and oriented in the direction shown.
Hidden lines are normally omitted.

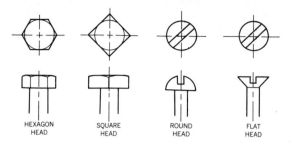

HEXAGON
HEAD

SQUARE
HEAD

ROUND
HEAD

FLAT
HEAD

Figure AI-62

Special holes ... often associated with fasteners.

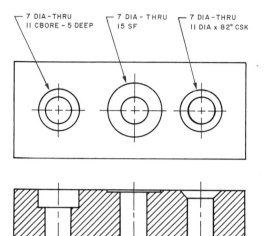

- 7 DIA – THRU
  II CBORE – 5 DEEP

- 7 DIA – THRU
  I5 SF

- 7 DIA – THRU
  II DIA x 82° CSK

COUNTERBORE   SPOTFACE   COUNTERSINK

Figure AI-63

AI.8 Intersections

The features of an object are composed of a series of intersecting geometric shapes.
The intersection lines may be straight or curved depending on the nature of the
intersecting surfaces.

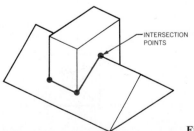

INTERSECTION
POINTS

Plane surfaces ... intersect in a straight line
that joins the points of intersection of lines of
one plane with the surface of the other plane.

Figure AI-64

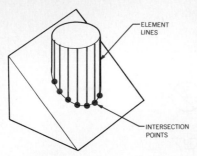

Curved surfaces and plane surfaces ... usually intersect in a curved line that joins the points where element lines on the curved surface intersect the surface of the plane.

Figure AI-65

Curved surfaces ... intersect in a curved line that joins the points where element lines on one curved surface intersect the other curved surface.

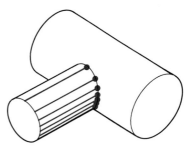

Figure AI-66

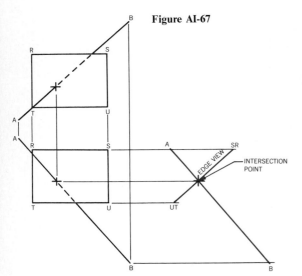

Figure AI-67

Line and a plane ... intersect in a point. This is an important step in locating intersection lines. It is also important in locating attachment points for support wires and points where wires pass through surfaces.

If an edge view of the plane is available or is easily drawn, the intersection point can be seen where the line passes through the edge view. The point can then be projected to other views.

If the edge view is not available, the intersection point can be found by passing the edge of a "cutting" plane through one view of the line. When the intersection line of the two planes is projected onto an adjacent view, it will cross the line at its intersection point.

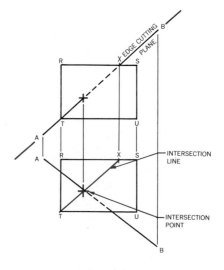

Figure AI-68

AI.9 Developments

Hollow objects that are made of thin material are often formed from one flat piece of material that is folded on "bend" lines and joined on "seam" lines. The development may be either an "inside" or an "outside" pattern depending on which side of the object is uppermost on the drawing.

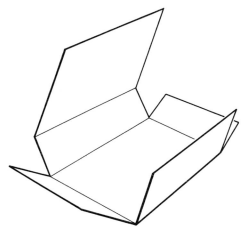

Figure AI-69

Prism development (open ends)

Draw two views of the prism . . . one showing the edge (bend) lines in their true length . . . the other showing the edge lines as points.

POINT VIEWS

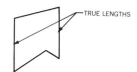
TRUE LENGTHS

Figure AI-70

Project all points of the prism perpendicular to the edge lines. Draw a "stretch-out" line parallel to the projection lines.

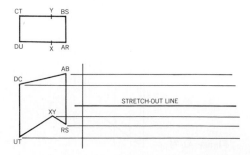

STRETCH-OUT LINE

Figure AI-71

Transfer the panel widths from the end view of the prism to the stretch-out line. Start at the desired seam line and proceed in the necessary order for an outside or an inside pattern.

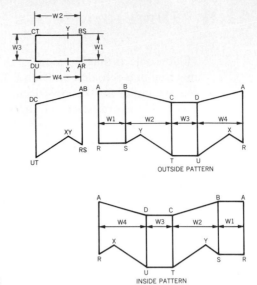

OUTSIDE PATTERN

INSIDE PATTERN

Figure AI-72

Cylinder development (open ends)

Draw two views of the cylinder . . . one showing the axis (centerline) in its true length . . . the other showing the axis as a point.

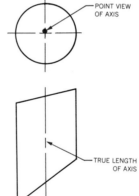

POINT VIEW OF AXIS

TRUE LENGTH OF AXIS

Figure AI-73

Draw equally spaced element lines on the surface parallel to the axis. These lines correspond to the edge (bend) lines of a prism.

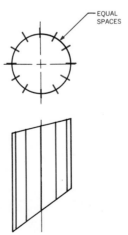

EQUAL SPACES

Figure AI-74

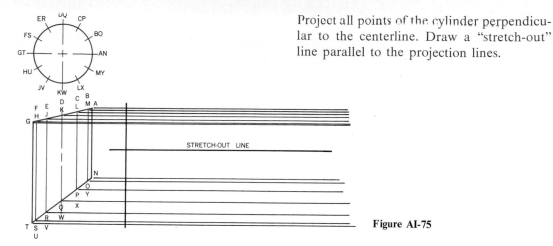

Project all points of the cylinder perpendicular to the centerline. Draw a "stretch-out" line parallel to the projection lines.

STRETCH-OUT LINE

Figure AI-75

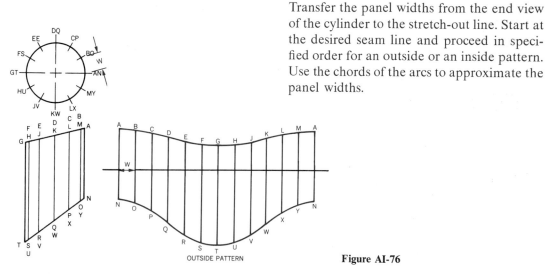

Transfer the panel widths from the end view of the cylinder to the stretch-out line. Start at the desired seam line and proceed in specified order for an outside or an inside pattern. Use the chords of the arcs to approximate the panel widths.

OUTSIDE PATTERN

Figure AI-76

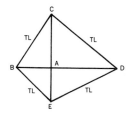

Pyramid development (open bottom)

Draw two or more views of the pyramid to show the true length of all its lines. Auxiliary views may be required to do this.

Figure AI-77

Starting at the desired seam line, lay out the true size of each triangular panel by swinging arcs with a compass set at the true length of each line. Follow in sequence in the specified order for an outside or an inside pattern.

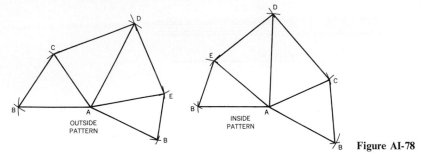

OUTSIDE PATTERN

INSIDE PATTERN

Figure AI-78

AI.10 Dimensioning

Before an object or a structure is manufactured or constructed, its shape is normally defined in a multiview drawing. The views on this drawing must also include all the dimensions necessary to define the exact size and location of the various features. These dimensions must be stated clearly and precisely with only one possible interpretation. For machine drawing, national and international standards have been established to aid in this process.

Defining shape

An object is composed of a series of geometric shapes the sizes and locations of which must be specified *only once* and in *only one way*.

Specify overall size (width, depth, and height) . . . in most cases.

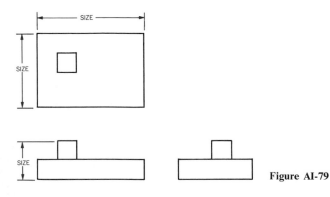

Figure AI-79

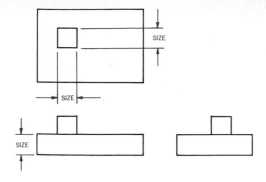

Specify sizes of individual shapes not already covered by overall dimensions. The function of the object determines which sizes to give. In this example, the height of the large block was judged to be more important than that of the small block.

Figure AI-80

Specify locations of the individual shapes. The functions of the object determines the surfaces that will be used for location.

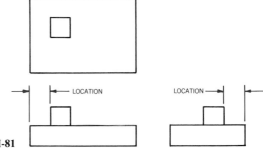

Figure AI-81

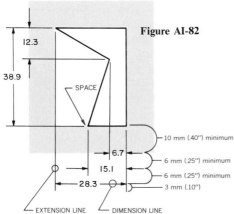

Figure AI-82

Place dimensions on the view where the contour of the geometric shape is most clearly seen. Avoid dimensions directed to hidden lines.

Dimension placement

Dimensions are placed outside the views whenever possible. Leave at least a 10 mm (0.40 in.) open space around each view before starting the dimension lines that show the numerical value of each dimension. The contours of the object are extended with extension lines that start with a small space and end 3 mm (0.12 in.) beyond the last dimension line. Extension lines may cross extension and object lines. Dimension lines may not be crossed.

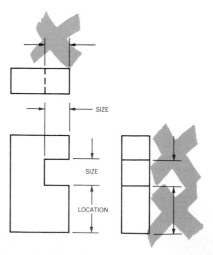

Figure AI-83

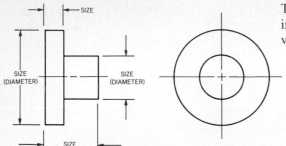

The size (diameter and length) of a solid cylinder is normally specified on its non-circular view.

Figure AI-84

The size (diameter and depth) of a hollow cylinder (a hole) is normally specified on its circular view using a leader and a note. The leader must aim at the center of the hole with the tip of the arrow touching the circle. The horizontal portion of the leader is at least 6 mm (0.25 in.) long.

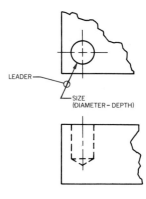

Figure AI-85

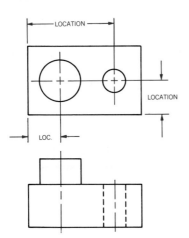

A cylindrical shape (solid or hollow) is located by its centerline in the circular view.

Figure AI-86

The size of a partial cylinder (a rounded corner) is specified by its radius in the circular view using a leader. The leader must aim at or pass through the center of the radius. The letter "R" must follow the numerical value of the size.

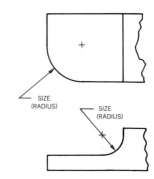

Figure AI-87

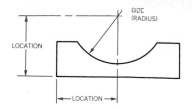

SIZE
(RADIUS)

LOCATION

LOCATION

If a partial cylinder is tangent to a flat surface, location dimensions for the center of its radius are not needed. If the partial cylinder is not tangent, the center of the radius must be located.

Figure AI-88

AI.11 Pictorial drawing

A pictorial drawing system is used when it is necessary to describe an object in only one view for general information purposes. Pictorial systems vary in their ease of use, their versatility in showing an object from different viewing directions, and their degree of realism in portraying the object. Hidden lines are normally omitted in pictorial drawing because most of the important features are visible if the viewing direction is properly selected.

Axonometric drawing

In multiview drawing each view normally shows only one face of the object.

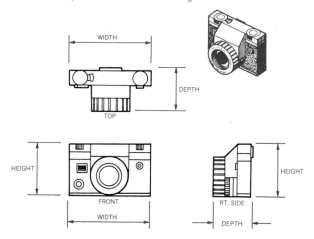

WIDTH

DEPTH

TOP

HEIGHT

FRONT

WIDTH

RT. SIDE

DEPTH

HEIGHT

Figure AI-89

Auxiliary views are required in multiview drawing to show two or three faces in one view. This procedure is time consuming, and the auxiliary view is not realistically oriented on the drawing paper.

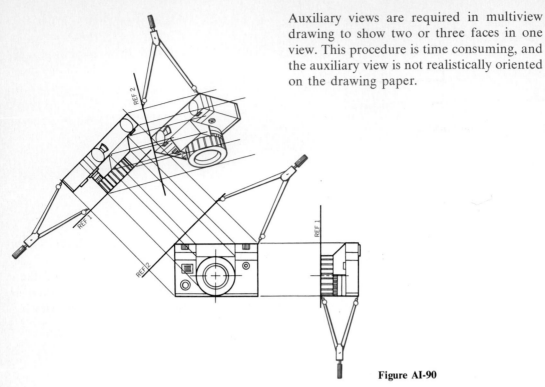

Figure AI-90

In axonometric drawing the same view as seen in the auxiliary view can be drawn directly and in a natural position on the paper. The basic line and shape relationships of plane surfaces shown in multiview drawing are retained. Parallel lines remain parallel. Circular features appear as ellipses.

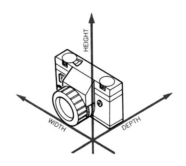

Figure AI-91

Axonometric drawing systems

In axonometric drawing, the object is positioned at an angle with the viewing plane. The three axonometric systems—isometric, dimetric, and trimetric—differ in the degree to which the three principal faces of the object are visible. This is done by increasing or decreasing the angle between each face and the viewing plane. As a face is tilted away from the viewing plane, its dimensions appear smaller. In drawing the view, this shortening is produced by using appropriate reducing scales on the width, height, and depth axes.

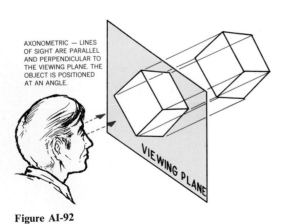

AXONOMETRIC — LINES OF SIGHT ARE PARALLEL AND PERPENDICULAR TO THE VIEWING PLANE. THE OBJECT IS POSITIONED AT AN ANGLE.

Figure AI-92

Isometric ... the three principal faces are equally visible because they are positioned at the same angle to the viewing plane. The width and depth axes are drawn at 30° to the horizontal. The same scale is used on all three axes because the dimensions are equally distorted.

In practice, to save time, the dimensions along all three isometric axes are transferred directly from the multiview drawing rather than by using a reducing scale. This makes the isometric system the easiest of the axonometric systems to use.

Figure AI-93

Dimetric ... two principal faces of the object are equally visible with the third face given greater or less visibility by varying its angle to the viewing plane. To obtain a better view of the top surface than in an isometric drawing, the object can be tilted up more. The width and depth scales remain equal, but the height scale is reduced.

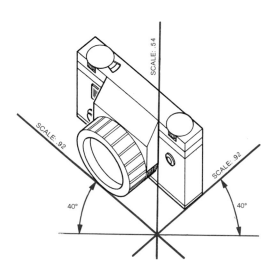

Figure AI-94

If the angle of tilt is less than in the isometric drawing, the top surface will be less visible. The width and depth scales remain equal, but the height scale is increased.

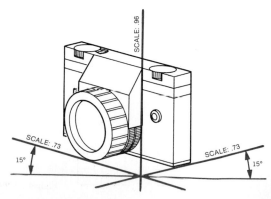

Figure AI-95

A better view of the front surface can be obtained by rotating the object so that the front is at a smaller angle to the viewing plane. In this case, the depth scale varies with the degree of rotation while the width and height scales remain equal.

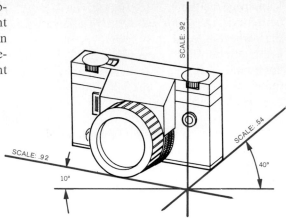

Figure AI-96

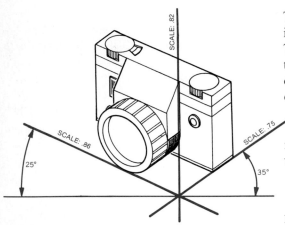

Trimetric . . . each principal face of the object is at a different angle to the viewing plane. This gives greater flexibility in positioning the three faces but is more time consuming to draw since different scales must be used on each axis.

Dimetric is more widely used than trimetric because it provides adequate views while requiring only two scales.

Figure AI-97

Layout of axonometric views

Set up the axis system desired with the required angles and scales.

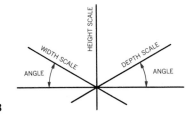

Figure AI-98

Imagine the object enclosed in a rectangular box that just touches its outermost points and surfaces.

Figure AI-99

Lay out the width, height, and depth of the enclosing box along the axonometric axes. Be sure to use the correct scale for each axis.

Figure AI-100

Next, draw in the surfaces of the object that fall inside the box. Locate the beginning and ending points of angled lines with measurements parallel to the width, height, and depth axes.

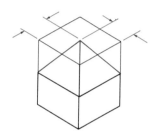

Figure AI-102

Perspective drawing

In perspective drawing more realistic pictorial views are produced. Although quite similar to the views produced in axonometric drawing, perspective drawing produces views that more closely duplicate how the human eye sees an object in space. Since the lines of sight are not perpendicular to the viewing plane, parallel lines on the object must be drawn so as to converge at an imaginary point in the distance. Width, height, and depth measurements must be made progressively smaller as they vanish into the distance. These variations from axonometric require additional time to draw accurately. This makes perspective drawing more suitable for artistic-type presentations.

On the outline of the enclosing box draw in the surfaces of the object that are in contact with the outside faces. Omit hidden lines.

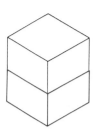

Figure AI-101

The circular opening on the right side of the object is drawn by laying out an enclosing square on the surface. The resulting rhombus serves as a guide for selection of an ellipse template opening that is tangent to the four sides.

A circle appearing on a surface is represented in axonometric by an ellipse the minor axis of which is parallel to the direction of a line perpendicular to the surface (the width axis in this example). A 35° ellipse template is used for circles on the principal faces of an object drawn in isometric.

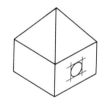

Figure AI-103

Figure AI-104

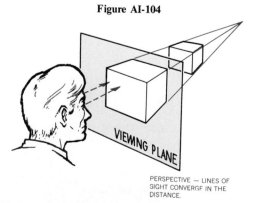

PERSPECTIVE — LINES OF SIGHT CONVERGE IN THE DISTANCE.

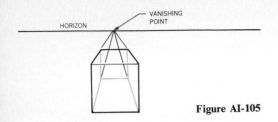

Figure AI-105

Perspective drawing systems. The three perspective systems differ in the direction in which the object is viewed and the degree of realism produced.

One-point perspective . . . object lines parallel to the depth axis converge to a single vanishing point on the horizon.

Two-point perspective . . . object lines parallel to the width and depth axes converge to two vanishing points on the horizon. This is the most commonly used perspective system.

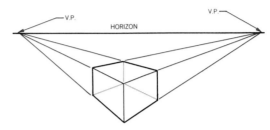

Figure AI-106

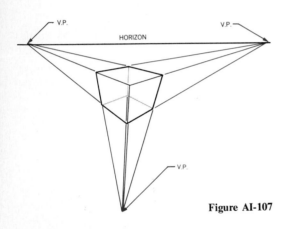

Figure AI-107

Three-point perspective . . . object lines parallel to width, depth, and height axes converge to three vanishing points.

Oblique drawing

In oblique drawing less realistic pictorial views are produced, but an acceptable view is often produced in less time. This makes the oblique system more useful for freehand sketching than for professional illustration.

A principal face of the object is positioned parallel to the viewing plane, thus giving this face and those surfaces parallel to it an undistorted appearance. Being true size and shape makes these surfaces easier to draw. The lines of sight are parallel but are at an angle to the viewing plane so that the other faces of the object can be seen.

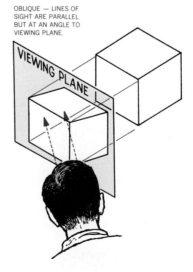

Figure AI-108

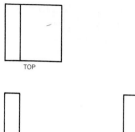

TOP

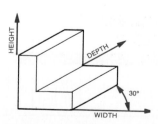

FRONT   RT. SIDE   **Figure AI-109**

Layout of oblique views Orient the object so that the face having the most contours is in the "front" viewing position and can be drawn undistorted. All measurements on or parallel to the width and height axes are the same as on the multiview drawing.

The depth axis is usually drawn at 45° to the horizontal. For convenience in sketching, depth measurements are usually made to the same scale as that used for width and height. Lines parallel to the depth axis remain parallel.

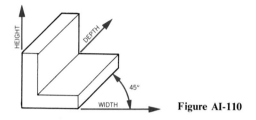

Figure AI-110

To create a more realistic appearance, depth measurements may be made to a reduced scale. The angle of the depth axis may also be changed and its scale changed accordingly.

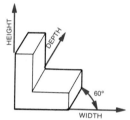

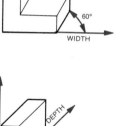

Figure AI-111

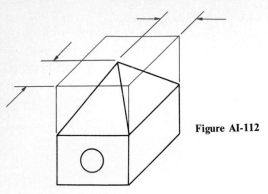

For objects that are basically rectangular, an enclosing box can be used to help in the layout. Orient the object so that circles and irregular contours appear on the undistorted front face.

Figure AI-112

For cylindrical objects, a longitudinal centerline is laid out first ... depth measurements are marked off on this axis to locate the centers of the circular faces.

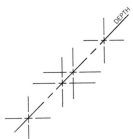
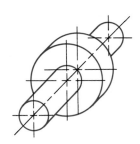

Figure AI-113

AI.12 Charts and graphs

Charts and graphs are used to portray graphically the relationship between various types of data. This comparison may be made using lines, areas, or volumes.

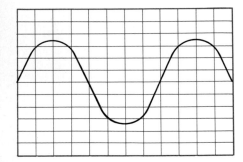

Graphs use lines to show the relationship between variables.

Figure AI-114

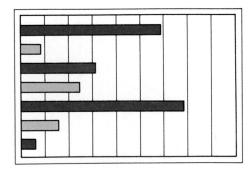

Figure AI-115

Bar charts use rectangular areas to compare several quantities.

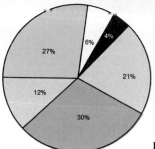

Circle ("pie") charts use segmented areas to show the various percentages that make up a whole.

Figure AI-116

Pictographic charts use pictures or symbols to show relationships.

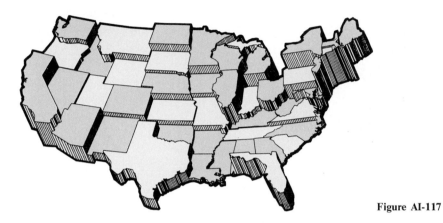

Figure AI-117

Preparing graphs

The quantities being compared are usually stated in order, with the dependent variable first and the independent variable second . . . the dependent is usually plotted on the vertical (ordinate) axis and the independent on the horizontal (abscissa) axis.

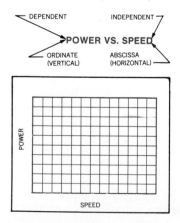

Figure AI-118

Preprinted graph paper makes the preparation of graphs much easier . . . scale divisions should be selected to produce a curve that will clearly display the functional relationship between the variables.

The scales are normally drawn inside the borders of the graph paper . . . scales must be labeled with the names of the variables and the units in which they are measured.

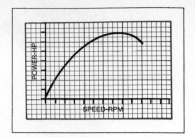

Figure AI-119

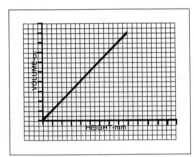

Figure AI-120

Plotted points are not usually shown on graphs of theoretical relationships.

Figure AI-121

When plotting experimental data, the points are normally enclosed in small circles . . . if more than one curve is plotted on the same sheet, other geometric shapes, such as triangles and squares, can be used for identification.

The curve is usually sketched in lightly to average the points and then darkening in using an irregular curve. The line starts and stops at the edge of each point symbol and need not pass through all points.

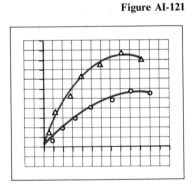

If a variable has a very large range of values, it can be compressed by plotting it on a logarithmic scale.

Semilog graph paper has uniform graduations on one axis and logarithmic graduations on the other.

Log-log paper has logarithmic graduations on both axes.

Figure AI-122

Graphs can also be plotted on polar coordinate graph paper.

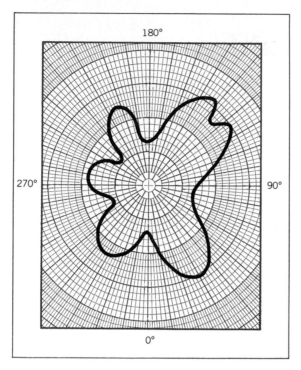

APPENDIX II

Conversion tables

The E minus or plus an exponent following a value in the table indicates the power of 10 by which this value should be multiplied. Thus, $6.214\,E - 06$ means 6.214×10^{-6}.

1. Length equivalents

cm	in.	ft	m	mi*	
1	$3.937\,E - 01$	$3.281\,E - 02$	$1.0\,E - 02$	$6.214\,E - 06$	cm
2.540	1	$8.333\,E - 02$	$2.54\,E - 02$	$1.578\,E - 05$	in.
$3.048\,E + 01$	$1.2\,E + 01$	1	$3.048\,E - 01$	$1.894\,E - 04$	ft
$1.0\,E + 02$	$3.937\,E + 01$	3.281	1	$6.214\,E - 04$	m
$1.609\,E + 05$	$6.336\,E + 04$	$5.280\,E + 03$	$1.609\,E + 03$	1	mi

* mile.

Additional Measures

Metric: 1 km = 10^3 m
 1 mm = 10^{-3} m
 1 μm = 10^{-6} m (micron)
 1 Å = 10^{-10} m (angstrom)

English: 1 mil = 10^{-3} in.
 1 yd = 3.0 ft
 1 rod = 5.5 yd = 16.5 ft
 1 furlong = 40 rod = 660 ft

2. Area equivalents

m²	in.²	ft²	acres	mi²	
1	1.55 E + 03	1.076 E + 01	2.471 E − 04	3.861 E − 07	m²
6.452 E − 04	1	6.944 E − 03	1.594 E − 07	2.491 E − 10	in.²
9.290 E − 02	1.44 E + 02	1	2.296 E − 05	3.587 E − 08	ft²
4.047 E + 03	6.273 E + 06	4.356 E + 04	1	1.562 E − 03	acres
2.590 E + 06	4.018 E + 09	2.788 E + 07	6.40 E + 02	1	mi²

Additional Measures

1 hectare = 10^4 m²

3. Volume equivalents

cm³	in.³	ft³	gal (U.S.)	
1	6.102 E − 02	3.532 E − 05	2.642 E − 04	cm³
1.639 E + 01	1	5.787 E − 04	4.329 E − 03	in.³
2.832 E + 04	1.728 E + 03	1	7.481	ft³
3.785 E + 03	2.31 E + 02	1.337 E − 01	1	gal (U.S.)

Additional Measures

Metric:
1 liter = 10^3 cm³
1 m³ = 10^6 cm³

English:
1 quart = 0.250 gal (U.S.)
1 bushel = 9.309 gal (U.S.)
1 barrel = 42 gal (U.S.)
  (petroleum measure only)
1 imperial gal = 1.20 gal (U.S.)
    approx.
1 board foot (wood) = 144 in.³
1 cord (wood) = 128 ft³

4. Mass equivalents

kg	slug	lb$_m$*	g	
1	6.85 E − 02	2.205	1.0 E + 03	kg
1.46 E + 01	1	3.22 E + 01	1.46 E + 04	slug
4.54 E − 01	3.11 E − 02	1	4.54 E + 02	lb$_m$
1.0 E − 03	6.85 E − 05	2.205 E − 03	1	g

* Not recommended.

5. Force equivalents

N*	lb$_f$†	dyn**	kg$_f$‡	g$_f$‡	Poundal‡	
1	2.248 E − 01	1.0 E + 05	1.019 E − 01	1.019 E + 02	7.234	N
4.448	1	4.448 E + 05	4.54 E − 01	4.54 E + 02	3.217 E + 01	lb$_f$
1.0 E − 05	2.248 E − 06	1	1.02 E − 06	1.02 E − 03	7.233 E − 05	dyn
9.807	2.205	9.807 E + 05	1	1.0	7.093 E + 01	kg$_f$
9.807 E − 03	2.205 E − 03	9.807 E + 02	1.0 E − 03	1	7.093 E − 02	g$_f$
1.382 E − 01	3.108 E − 02	1.383 E + 04	1.410 E − 02	1.410 E + 01	1	poundal

* Newton.
† Avoirdupois.
** Dyne.
‡ Not recommended.

6. Velocity and acceleration equivalents

Velocity

cm/s	ft/s	ml/h (mph)	km/h	
1	3.281 E − 02	2.237 E − 02	3.60 E − 02	cm/s
3.048 E + 01	1	6.818 E − 01	1.097	ft/s
4.470 E + 01	1.467	1	1.609	mi/h
2.778 E + 01	9.113 E − 01	6.214 E − 01	1	km/h

Acceleration

cm/s²	ft/s²	ḡ*	
1	3.281 E − 02	1.019 E − 03	cm/s²
3.048 E + 01	1	3.109 E − 03	ft/s²
9.807 E + 02	3.217 E + 01	1	ḡ

*Standard acceleration of gravity.

Additional measures

1 knot = 1.152 mi/hr

7. Pressure equivalents

Pressure

dyn/cm²	N/m² pascal	lb/in.² (psi)	lb$_f$/ft² (psf)	atm*	Head† in. (Hg)	ft (H$_2$O)	
1	1.0 E − 01	1.45 E − 05	2.089 E − 03	9.869 E − 07	2.953 E − 05	3.349 E − 05	dyn/cm²
1.0 E + 01	1	1.45 E − 04	2.089 E − 02	9.869 E − 06	2.953 E − 04	3.349 E − 04	N/m²
6.895 E − 04	6.895 E + 03	1	1.44 E + 02	6.805 E − 02	2.036	2.309	lb$_f$/in.²
4.788 E + 02	4.788 E + 01	6.944 E − 03	1	4.725 E − 04	1.414 E − 02	1.603 E − 02	lb/ft²
1.013 E + 06	1.013 E + 05	1.47 E + 01	2.116 E + 03	1	2.992 E + 01	3.393 E + 01	atm
3.336 E + 04	3.386 E + 03	4.912 E − 01	7.073 E + 01	3.342 E − 02	1	1.134	in (Hg)
2.986 E + 04	2.986 E + 03	4.331 E − 01	6.237 E + 01	2.947 E − 02	8.819 E − 01	1	ft (H$_2$O)

*Standard atmospheric pressure.
†At std. gravity and 0°C for Hg, 15°C for H$_2$O.

Additional Measures

1 bar = 1 dyne/cm²
1 pascal = 1 N/m²

8. Work and energy equivalents

J*	ft-lb$_f$	W-h	Btu†	Kcal**	kg-m	
1	7.376 E − 01	2.778 E − 04	9.478 E − 04	2.388 E − 04	1.020 E − 01	J
1.356	1	3.766 E − 04	1.285 E − 03	3.238 E − 04	1.383 E − 01	ft-lb$_f$
3.60 E + 03	2.655 E + 03	1	3.412	8.599 E − 01	3.671 E + 02	W-h
1.055 E + 03	7.782 E + 02	2.931 E − 01	1	2.520 E − 01	1.076 E + 02	Btu
4.187 E + 03	3.088 E + 03	1.163	3.968	1	4.269 E + 02	Kcal
9.807	7.233	2.724 E − 03	9.295 E − 03	2.342 E − 03	1	kg-m

* Joule.
† British thermal unit.
** Kilocalorie.

Additional measures

1 Newton-meter = 1 J
1 erg = 1 dyne-cm = 10^{-7} J
1 cal = 10^{-3} kcal
1 therm = 10^{-5} Btu
1 million electron volts (Mev) = $1.602(10^{-13})$ J

9. Power equivalents

J/s	ft-lb$_f$/s	hp*	kW	Btu/h	
1	7.376 E − 01	1.341 E − 03	1.0 E − 03	3.412	J/s
1.356	1	1.818 E − 03	1.356 E − 03	4.626	ft-lb$_f$/s
7.457 E + 02	5.50 E + 02	1	7.457 − 01	2.545 E + 03	hp
1.0 E + 03	7.376 E + 02	1.341	1	3.412 E + 03	kW
2.931 E − 01	2.162 E − 01	3.930 E − 04	2.931 E − 04	1	Btu/h

* Horsepower.

Additional Measures

1 W = 10^{-3} kW
1 cal/s = 14.29 Btu/h
1 poncelet = 100 kg-m/sec = 0.9807 kW
1 ton of refrigeration = 1.2×10^4 Btu/h

Time

1 week	7 days	168 hours	10,080 minutes	604,800 seconds
1 mean solar day		24 hours	1440 minutes	86,400 seconds

1 calendar year 365 days 8760 hours 5.256(E + 05) minutes 3.1536(E + 07) seconds
1 tropical mean solar year   365.2422 days (basis of modern calendar)

Temperature

$\triangle 1°$ Celsius (formerly Centigrade)   (C) = $\triangle 1°$ Kelvin (K) = 1.8° Fahrenheit (F)
= 1.8° Rankine (R)

0°C = 273.15°K = 32°F = 491.67°R = 0°R $°C = \frac{5}{9}(°F - 32)$
0°K = −273.15°C = −459.67°F

Electrical

$°F = \frac{9}{5}(°C + 32)$

1 coulomb	$1.036(10)^5$ faradays	0.1 abcoulomb	$2.998(10)^9$ statcoulombs
1 ampere		0.1 abampere	$2.998(10)^9$ statcoulombs
1 volt	10^3 millivolts	10^8 abvolts	$3.335(10)^{-3}$ statvolt
1 ohm	10^6 megohms	10^9 abohms	$1.112(10)^{-12}$ statohm
1 farad	10^6 microfarads	10^{-9} abfarads	$8.987(10)^{11}$ statfarads

Index